Astronomers' Universe

For further volumes:
http://www.springer.com/series/6960

George Rhee

Cosmic Dawn

The Search for the First Stars and Galaxies

 Springer

George Rhee
Department of Physics & Astronomy
University of Nevada
Las Vegas, Nevada, USA

Cover image credits:

Simulation: Matthew Turk, Tom Abel & Brian O'Shea
Visualization: Ralf Kaehler and Tom Abel (KIPAC/Stanford)

ISSN 1614-659X
ISBN 978-1-4614-7812-6   ISBN 978-1-4614-7813-3 (eBook)
DOI 10.1007/978-1-4614-7813-3
Springer New York Heidelberg Dordrecht London

Library of Congress Control Number: 2013940914

Printed on acid-free paper

Springer is part of Springer Science+Business Media (www.springer.com)

To Heidie,
  to the memory of Albert and Helen Rhee

De Guiche	Monsieur, Have you read *Don Quixote*?
Cyrano	Read it? I've practically lived it.
De Guiche	I suggest you study the passage about the windmills.
Cyrano	Chapter thirteen.
De Guiche	When you make war on windmills you may find that the mill sails will swing their heavy spars and cast you down into the mud.
Cyrano	Or up among the stars!

Edmond Rostand. Cyrano de Bergerac.

Preface

Ulysses is telling Dante about his final fatal voyage, beyond the Pillars of Hercules, beyond the borders of the known world. Thanks to this episode in Dante's poem, Ulysses has become one of the worlds great symbols of human dignity and human resource, a representative of the human compulsion to follow knowledge ... Ulysses goes on to tell Dante of the courage that was required to initiate and pursue his adventure; I set forth then upon the open sea with just one vessel from my fleet's remains and those few men who had not deserted me.

Seamus Heney, Speech delivered to the Human Rights Organization
Frontline January 2002

When I was a boy I attended lectures by Rafel Carreras at the European Center for Nuclear Research (CERN) in Geneva Switzerland. These weekly popular science lectures were attended by people from all walks of life. As Carreras puts it

> I often had university professors in the audience, schoolchildren, pensioners, sometimes even a mother feeding her baby, not forgetting the former CERN staff member whose job in my early years had been to clean off the blackboards at the end of my lectures, who, after he had retired, used to attend my lectures and sit in the front row

Carreras' job description at CERN was to "contribute to the intellectual health of the staff". Maybe every organization should have such a position. I hope this book will fill the reader with enthusiasm and encourage them to further explore the wonderful field of astronomy. It is in the Carreras spirit of sharing knowledge that I have written this book.

The book is aimed at the general public. It is not intended to be a textbook but rather an accessible overview of cosmology. The purpose of this book is to guide the reader to one of the frontiers of the field; the search for the first galaxies that formed after the big bang.

I feel I have two main duties as an author in writing a book such as this. Firstly, the book should not be boring, and secondly, it should provide correct scientific information. I hope I have achieved the goal of being entertaining without compromising accuracy.

Progress in cosmology is driven on the one hand by creative thinking and on the other by vast improvements in computing, telescope and instrument design. We stand on the brink of uncharted territory to be explored with the next generation of telescopes. It is easy for astronomers to impress the public with the awe inspiring images produced by telescopes such as the Hubble Space Telescope. I hope after reading this book, the reader will have the insight to appreciate the deeper significance of these images placed in their scientific context. For the most part I hope the reader will feel the same sense of wonder that I felt as a boy when I discovered science. Knowledge and wonder go hand in hand in the field of cosmology.

I would like to thank my colleagues at the University of Nevada, Stephen Lepp, Tao Pang, Bing Zhang, Daniel Proga and Ken Nagamine for providing a stimulating intellectual environment in which to think and write about these problems. Colleagues at other institutions including Rien van de Weygaert, Anatoly Klypin, Octavio Valenzuela and Fabio Governato have helped shape my ideas. I acknowledge support by the NSF AST-04-07072 grant to the University of Nevada, Las Vegas. I thank Doug Haag for his careful reading of the manuscript. I thank Heidie Grigg for her constant love and encouragement that made it possible for me to do this work. Let us then, like Ulysses, set forth on the open sea...

Las Vegas, NV, USA   George Rhee

Contents

x   Contents

Abbreviations

I have tried to avoid abbreviations as much as possible in this book, but a few have crept in that I list below.

ALMA   Atacama Large Millimeter Array
CCD   Charge Couple Device
CERN   European Center for nuclear Research
COBE   Cosmic Background Explorer
ESO   European Southern Observatory
ESA   European Space Agency
E-ELT ESO-Extremely Large Telescope
HST   Hubble Space Telescope
NASA   National Aeronautics and Space Administration
JWST   James Webb Space Telescope
SDSS   Sloan Digital Sky Survey
SKA   Square Kilometer Array
VLA   Very Large Array
WMAP   Wilkinson Microwave Anisotropy Probe

Part I
Prologue

In this section we introduce ideas that form the basis of cosmology. Chapter 1 is a review of the historical development of some of these ideas. Chapter 2 presents the three main pieces of evidence for the Big Bang theory that explains the early development of the universe. Chapter 3 describes observations of stars and galaxies. Chapter 4 presents the main evidence for dark matter which we believe is the dominant form of matter in the universe.

This sets the stage for the main topics of the book. Part II explains how we think galaxies and the cosmic web that they constitute came into existence. Part III describes our attempts to observe the dawn of the universe using telescopes currently being designed and built. This search is expected to discover the first stars and galaxies to form after the big bang.

1. Cosmology Through Its Past

The Greeks were the first mathematicians who are still 'real' to us today... So Greek mathematics is permanent, more permanent even than Greek literature. Archimedes will be remembered when Aeschylus is forgotten because languages die and mathematical ideas do not.

G.H. Hardy, A Mathematician's Apology

One of the characters in the 1950s British comedy radio series, The Goon Show, once remarked that "Everybody's got to be somewhere." The answer to the question of where we are in the universe and how we got there has changed dramatically over the past 2,000 years. It is a question that all cultures try to answer in one way or another. We discuss in this chapter the history of our attempts to answer this question. I begin with a Native American myth and then discuss Greek thought and the idea of rational inquiry. The ideas of motion in the solar system are discussed leading to the work of Isaac Newton. This, in turn, leads to thoughts on cosmology and the infinite universe. The telescope enters the stage, and we discuss its use in changing our view of the solar system and how galaxies was discovered. We close this chapter with the story of how the nature of galaxies was revealed and a description of work on cosmology in the first half of the twentieth century.

A Journey Back in Time: The Grand Canyon

The Grand Canyon of the Colorado River is one of the most spectacular places on earth. To journey into the Grand Canyon is to take a trip into the Earth's past. At the deepest point the rocks one sees are over 2 billion years old, one-seventh of the age of the

G. Rhee, *Cosmic Dawn: The Search for the First Stars and Galaxies*,
Astronomers' Universe, DOI 10.1007/978-1-4614-7813-3_1,
© Springer Science+Business Media, LLC 2013

universe. One of many striking places in the Grand Canyon is the confluence of the Little Colorado River with the main Colorado River. Here, one can see a major part of the Earth's history at a glance, from the 300-million-year-old rocks at the rim to the 2-billion-year-old rocks at the bottom. Near this place is a mound of earth that has great significance for the native Hopi people, who believe that the first people were created in a cave deep below the Earth's surface. According to the Hopi myth, these first people climbed up through caves from lower worlds until they reached the earth's surface and entered the fourth world through a hole in the earth known as the Sipapu. For the Hopi this is the holiest spot on earth. The Hopi people believe that the fourth world is sacred and that if the land is abused they will lose their sacred way of life.

This myth speaks to the human thirst for knowledge of origins. Our sense of identity is linked to our sense of history. Cosmology seeks to answer the same questions that myths address. How did things come to be the way they are today? Where do we come from? How did we get here? The history of cosmology reveals that our answers are determined by our view of the universe and include the limitations of that view. Our cosmology is determined by how much of the universe we can see with our eyes and our telescopes, and, for most of history, astronomy was done without telescopes.

We shall see that our current questions are more specific; What were the first objects to light up the universe and when did they do it? How do cosmic structures form and evolve? What are dark matter and dark energy?

Magic Reason and Experience: The Legacy of the Greek Thinkers

Our discussion of the origin of modern science begins with the works of ancient Greek thinkers. It is generally agreed that inquiries that are recognizable as science and philosophy were developed in the ancient world. Important developments took place from the sixth to the fourth centuries B.C. It is in this period that writers began to criticize what they called magical beliefs, and in particular criticized claims of the ability to forcibly manipulate the divine or supernatural. As an example, a treatise was written that exposed as frauds those who claimed to be able to cure epilepsy by purification, incantations, and other rituals.

The Greek idea of astronomy involved using a model to reproduce the observed motions of celestial objects in the night sky. Most Greek cosmologists placed the earth at the center of the universe. To our modern eyes, this seems perhaps egotistical. What is so important about our little planet that it should be at at the center of the universe? Greek scientists had actually searched for evidence of the Earth's motion through space and found it lacking. Placing the earth at rest at the center was the simplest hypothesis consistent with the available facts.

Greek astronomers noticed that stars retain their positions relative to each other from night to night. The shape and relative positions of the constellations do not appear to change from one year to the next. This was true for all but five stars which appeared to move from one constellation to the next throughout the year. They called these objects planets, the Greek word for wanderers. The Greeks studied the motions of these planets relative to the stars and noticed that at certain times of the year a planet would stop its drift relative to the stars and change direction for a few weeks, then reverse its direction again, a phenomenon known as retrograde motion.

As we shall see, retrograde motion is really an optical illusion caused by the way the Earth passes other planets on its orbit around the Sun. Greek astronomy consisted of the study of the solar system, but their mathematical approach to the problem has enabled scientists to develop the big bang theory.

Mathematics, the Language of the Book of Nature

The Greek universe consisted of planets moving against a backdrop of fixed stars. The central issue for Greek cosmology was thus to explain planetary motion. The problem of planetary motion could have been solved in an easy way by invoking spirits. According to that view, the planets start their backwards motion during a certain month because they feel like it. It was not known at the time that planets are inanimate pieces of rock or gaseous spheres. When I walk down the street and suddenly stop, realizing

I have not locked the house, I turn around of my own free will and go back and lock it, perhaps planets behave the same way. The Greeks had the deep insight to formulate a question that was to yield a fruitful answer. Might there not, they asked, be simple mathematical laws that govern motion of the planets? It turns out, remarkably, that the laws of physics can be written in mathematical form. Why this should be is a deep mystery. The world around us is remarkably complex. Weather patterns, the ebb and flow of life, human interactions, cannot be quantified by simple laws. Yet surprisingly, underlying all this complexity are simple laws that govern the behavior of all matter.

A hydrogen atom at the other end of our galaxy, 100,000 light years away will have exactly the same properties as a hydrogen atom in my body. This is a reflection of the fundamental laws of physics. The Greeks did not discover the fundamental laws of nature or even have the correct answers regarding motion in our solar system. Much more importantly however, they were asking the right questions. Science is the art of the soluble. The question the Greek astronomers posed was "is it possible to construct a simple mathematical model that explains the observed motions of the planets?" The model they constructed, much like our physical laws today, was an approximation, that provided a good fit to the best available measurements of the time.

The Greeks believed that the heavens embody perfection. The most perfect mathematical object known to them was a circle, so the Greeks constructed models based on uniform circular motion. The Greek model of the cosmos consisted of concentric spheres with the Earth at the center. To account for retrograde motion it was necessary to have at least two spheres for each planet. In practice a model for the solar system could consist of 20–50 spheres. For each planet a large sphere was necessary to account for the general drift of a planet and a smaller sphere was required to account for retrograde motion in the manner that we observe it. The model was described by Ptolemy (circ. AD 100–170) in a book called the Almagest which explained the motions of heavenly bodies and gave instructions as to how to calculate them. The Ptolemaic system agreed quite well with the observations available at the time and remained in use for the next 1,500 years. As I will explain in more detail below, astronomical observations made in the sixteenth and seventeenth centuries ultimately disproved the model.

Our modern view is that the planets orbit the Sun in elliptical orbits because of the gravitational attraction between them and the Sun. Why did it take hundreds of years to find the correct answer? It took a long time to develop the necessary tools. Tycho's measuring instruments and the telescope used by Galileo as well as the mathematical tool of the calculus invented by Newton, were the key innovations that made it possible to solve the riddle of planetary motion.

This is a very difficult problem to solve. Imagine you are in the dark on a merry-go-round inside a big tent. On the inside of the tent someone has put lights that twinkle like stars. The merry-go-round is built in a complicated way with arms that can swivel in various directions. Five of your friends are riding on the merry-go-round and they have each been given a small candle to hold in their hands. You then have to deduce from the motion of the candles against the backdrop of stars how the merry-go-round is constructed. This is not an easy task! You have one big advantage over astronomers however. You can feel the air go by when you move, so you know at least that you are in motion. We cannot feel the motion of the Earth, even though modern astronomers believe the Earth is moving at about 200 km per second around the center of our galaxy. Since we feel nothing, it seems natural to assume that we are not moving. The Greeks, as we have mentioned, had considered and rejected the possibility that the Earth does revolve around the Sun.

There is an effect known as parallax, which one would expect to observe due to the Earth's motion around the Sun. The effect is easily demonstrated. Hold your hand extended at arm's length in front of you with one finger raised pointing upwards. If you close your right eye and view your finger with your left eye, then close your left eye and view your finger with your right eye, you will notice that your finger will appear to have moved relative to background objects, say at the other end of the room. Now move your finger close to your face, say a foot away, and repeat the experiment. Your finger when held closer to your face will appear to move even more. The object of this exercise is to demonstrate that you can judge the distance of an object (such as your finger) by seeing how much it appears to move when viewed against a distant background (in this case, the back of the room) from two vantage points (your left and right eye). Suppose that at the back of the room there are many stars and that your finger represents a foreground star. Your left eye would represent the position of the Earth in July and right eye the position of the Earth in January. As your finger did in the experiment you just performed, we would expect the foreground star to appear to move relative to the more distant stars when viewed from earth at a 6-month interval.

The Greeks looked for this effect and did not find it, concluding that the Earth is at rest at the center of the universe. This is a great example of the scientific method: You make a hypothesis, which implies a certain outcome for an experiment and perform the experiment. The experimental result does not conform with the prediction and the hypothesis is falsified. This is the method by which the Greeks established that the Earth is at rest. Why didn't the Greeks observe a parallax?

The parallax is in fact a measurable effect. It was measured for the first time in 1838 by Friedrich Bessel, a German astronomer and mathematician. He measured the parallax of a star called 61 Cygni and determined its distance from the Earth to be 11 light years. This is a huge distance by the way. The Sun is only 8 light minutes away from the Earth. Distances in astronomy are often measured in terms of the distance that light travels in a certain amount of time. A light year, the distance that light travels in a year, is 10^{16} m. One followed by 16 zeros or about 6 trillion miles.

It thus comes as no surprise that the angular shift that Bessel measured was very small, in fact about one six-thousandth of a moon diameter. The observational accuracy at the time of Hipparchus (c. 190–125 BC) was about one moon diameter. Tycho Brahe (1546–1601) built instruments that improved the accuracy of Hipparchus by a factor of 30; still far short of what was required to detect parallax. It is not possible to make such a measurement without the aid of a telescope. The Greeks had failed to discover parallax because they lacked the technology to measure the positions of stars in the sky to the required accuracy.

When one makes a scientific measurement to determine a small quantity (such as parallax, for instance) and finds nothing, one usually places an upper limit on the result. Instead of saying "the displacement is zero" one says "the displacement is less than some specified value," which in turn means that the star must be farther away than some specified distance. The same issue arose in the 1980s with the study of neutrino mass. Some scientists thought that since the mass of the neutrino had to be less than a very small number, it was zero. There is now clear evidence that neutrinos have mass. Such dubious lines of reasoning are still present in modern science.

The size of the parallax effect measured by Bessel had important implications. It meant that the distance to the nearest star is much larger than the distance from the earth to the Sun. To put it another way, if we were to shrink the solar system so that the Earth was only nine feet from the Sun, the entire solar system would be the size of a football field and the nearest star would be over 200 miles away. It was hard for Greek scientists to conceive how distant from us the stars really are. To truly mimic the parallax effect with your finger, you would have to hold it 30 or so miles from your face.

The Greek philosopher-scientists were placed on a pedestal and treated as great authorities by scholars working in the Middle Ages. During this period, scientific discussions centered not on nature itself but on Aristotle's opinion of nature. Science very often progresses thanks to new experiments and improved accuracy in observations. If one turns away from experimental studies and believes that the answers are all to be found by studying books alone, one is on slippery ground. The cult of personality exists in many fields of human endeavor from politics to sports and the arts. To come under the spell of a powerful personality can be great in the development of a person as long as it does not last too long. When this happens to a community of scholars it can stall progress. One can't question everything all the time but it is good to take some degree of responsibility for the opinions one holds to be true. The key development for astronomy in the middle ages was the construction of precision instruments for measuring angles on the sky and a quantitative understanding of the accuracy of the measurements. With the new measurements made with these new instruments it eventually became clear that all models involving circular orbits were obsolete.

Scientists forge new insights and create new theories because they do not accept the conventional wisdom and because they have access to new tools.

The iconic twentieth century physicists Einstein and Feynman constructed theories using deep physical intuition combined with mastery of mathematical tools. In the field of biology, Francis

Crick and James Watson used the new tool of X-ray diffraction to discover the structure of the DNA molecule.

World in Motion

The Greeks had developed an earth-centered model of the universe consisting of nested spheres rotating at uniform speed. This view was to be seriously challenged by a number of scholars in the sixteenth century. Men such as Copernicus, Kepler, Galileo, Tycho, and Newton succeeded in taking what was good about the past and building on it. A mix of tradition and innovation is the key to success in science. As we shall see, these men were working in a very different intellectual climate from that which prevails today. One could be burned at the stake for holding opinions that conflicted with official teachings. In our modern Western world, scientists may be jailed for disseminating military secrets, such as encryption codes, but no one gets persecuted by the government for their astronomical opinions. Climate science may be an unfortunate exception to this rule.

In the Middle Ages we see the development of what is recognizably a university with students working toward degrees in places like Bologna, Paris, and Oxford. Scholars moved from one center of learning to the next, and criticism of the Greek geocentric system began to emerge.

A standard astronomical text of the time was the *Tractatus de Sphaera* by Johannes de Sacrobosco written around 1230, which gave a simplified description of the standard model of Ptolemy. The totality of the universe was a spherical earth at the center of the solar system with stars placed in a spherical shell beyond Saturn. Nikolaus de Cusa known as Cusanus published in 1404 a book entitled 'On Learned Ignorance'. Scientists are not searching for the absolute truth but developing ideas or theories that are increasingly closer approximations to the truth. For this process to work one must make repeatable measurements to the best possible accuracy. Cusanus also stated that the Earth was not at the center of the universe and not at rest. He challenged medieval wisdom and set the stage for thinkers such as Giordano Bruno (1548–1600). Bruno believed in the modern idea of an infinite universe with the Sun as just an ordinary star among many.

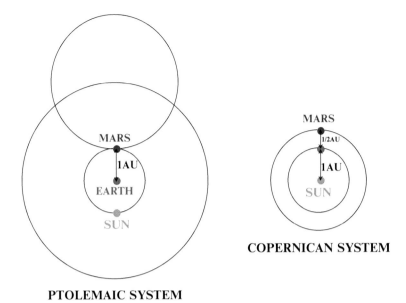

PTOLEMAIC SYSTEM

COPERNICAN SYSTEM

FIG. 1.1 A comparison of the orbit of Mars in the Earth centered model of Ptolemy (*left figure*) and the Sun centered model of Copernicus (*right figure*). In the Copernican system the minimum Mars-Earth distance is $\frac{1}{2}$ AU whereas in the Ptolemaic system it is 1 AU. AU stands for astronomical unit. 1 AU is the average distance between the earth and the Sun, a useful unit of measure for the solar system

A Polish monk, Nicolaus Copernicus (1473–1543), studied the Sun-centered system in quite some detail. Copernicus was not employed as a professor when he came up with his original idea concerning a Sun-centered, or heliocentric model of the universe (similarly, Einstein developed some of his theories while working in a patent office in Bern Switzerland). From the Sun-centered or heliocentric theory, a number of points emerged (Fig. 1.1).

First, in the heliocentric model, there is a natural explanation for why Mercury and Venus are never seen very far from the Sun in the sky. This is because in Copernicus' model they are close to the Sun in space. In the geocentric model this observation has no natural explanation and must be put into the model in an ad hoc manner. Another key point is that retrograde motion emerges as an optical illusion in the Sun-centered model. In the geocentric model the planets must stop in their tracks and reverse path, while in the Sun centered model it only appears that way, as viewed from earth. The planets in the Copernican system all circle the Sun and only appear to move backwards when the earth is overtaking them in their orbits. We can view the solar system as a racetrack, with the inner planets moving faster in space than the outer planets. Once in a while the Earth will catch up with a planet and overtake it and that planet will appear to move backwards in the sky as seen from earth. The Moon, however, really is orbiting the Earth and so never appears to undergo retrograde motion. A strength of the Copernican system is that it accounts for the motions of the planets in a simpler way than the geocentric system. It is, as scientists say, more elegant. Copernicus was not sure whether the universe was infinite or finite, he was aware of the absurdity of the rotation of an infinite universe surrounding the Earth, implying increasing rotation speeds for the more distant stars. It seems in the end that Copernicus placed the stars in a shell beyond Saturn. Thomas Digges (c. 1546–1595) published a version of the Copernican model that includes stars distributed in an unbounded universe.

In practice, the Copernican system was just as cumbersome as the geocentric system in that it required just as many spheres. One reason for this is that, although the planets do in fact circle the Sun, they do not move at uniform speed around the Sun. The Copernican model was also quite inaccurate, being off by as much as ten moon diameters in its prediction of planetary positions in the sky. Setting the Earth in motion raised a number of major problems. If the Earth is in motion, why do we not feel it? When we travel by car, we feel a jolting motion as we accelerate and we feel the bumps in the road. We also have to explain how the Earth goes around the Sun and does not leave the Moon behind.

A formidable but less rational obstacle for the Copernican theory was raised by the fact that the theory contradicted the teachings of the Bible and hence challenged the authority of the church. Copernicus died in 1543 and his book *On the Revolution of the Heavenly Spheres* was published that same year. Looking back at the sixteenth century from our present-day era, this is striking. These days, scientists rush into press, eager to claim credit for any idea. Copernicus only consented to publication of his theory when he was on his deathbed. Through the publication of his book his name lives forever in the history of science. It is striking that the preface to the first edition contains a disclaimer that would make any lawyer proud. It states essentially that the material in the book is pure and idle speculation and should not be taken literally. The preface implies that anyone who takes the contents of the book seriously is a fool. It turned out that the preface was inserted by a follower of Copernicus who wanted to avoid trouble with church's authorities. Johannes Kepler, whom we are about to meet, was enraged by this and wrote a critical letter stating that Copernicus *did* in fact mean what he wrote. The fact that one could come to serious harm for challenging the church's authority surely explains Copernicus' hesitation to publish. In the long run, history shows that ideas flourish best in an open society. Attempts to control the flow of knowledge in a totalitarian manner are bound to failure. One can force people to learn nonsense through threats, but one cannot dictate scientific truth from a position of political authority.

Thus far we have described two competing theories of the universe, neither of which is completely satisfying. For further progress on this problem better data were required. The next major actor in our story, Tycho Brahe (1546–1601), provided these data. In terms of modern science, Tycho (traditionally, he is referred to by his first name) strikes me as a politician as well as a researcher. A theorist such as Copernicus needed a pencil, paper, and wastebasket to do his work, as well as a fine mind. To obtain the best set of planetary observations ever made, Tycho needed the sixteenth-century equivalent of a modern research institute, which in turn required funding. He thus needed powers of persuasion to explain to non-scientists who controlled the purse strings why this work was important. It has been suggested that Tycho's observatory cost a few percent of the income of the King of Denmark. NASA costs a similar fraction of government spending today.

American scientists needed similar skills to Tycho's when they went to Congress to start lobbying for an optical telescope to be put in space. But let us not carry the analogy too far. Tycho's institute included dancing bears and dwarves for entertainment but no bears have been spotted at the Space Telescope Science Institute in recent times. Tycho obtained funding from the King of Denmark and set up his research institute on the island of Hveen. He then proceeded to accumulate a database of planetary positions over the next 20 years. Tycho took into account measuring errors, including those due to atmospheric refraction and the flexing of his instruments when they were pointed at different angles in the sky. One of his instruments is shown in Fig. 1.2.

FIG. 1.2 One of the sophisticated instruments that Tycho used to make his measurements. This instrument was used to measure the angular position of stars above or below the celestial equator. The *large circle* has a diameter of 2.7 m. This instrument was completed in 1585 (Published in 1662 and based on Tycho's original version, this illustration is by Willem Blaeu, a former assistant at Tycho's Hven observatory. Credit: Collection of Owen Gingerich)

Tycho made an amazing discovery in 1572. He observed the appearance of a new star in the constellation Cassiopeia. We now know that this was an exploding star, or supernova. The star blew itself apart and today over 400 years later we can see an expanding shell of gas where the star once was. The gas is so hot that it glows at X-ray wavelengths. Supernovae play a key role in cosmology, we shall meet them again later in this book. The fact that a star could dramatically change in brightness over a period of mere days was in direct contradiction with Greek cosmology, which stated that the heavens are unchanging. What if the star was not a star but a nearby disturbance? Tycho noted that it showed no parallax, that it seemed fixed in a given constellation. This discovery illustrates a wonderful aspect of science: the unexpected. You set out to do one thing only to find something completely different. In more modern times, radio astronomers set out to study the trails of meteors in the upper atmosphere and discovered very distant galaxies using their radio telescopes.

Tycho constructed a theoretical model that had the planets circling the Sun, which, in turn, circled the Earth. From our present vantage point this model seems contrived, but it was motivated by Tycho's failure to detect parallax. Eventually, as happens in the best of careers, there was a change of king, and Tycho did not get on with the new king. Tycho eventually left Denmark and ended up in Prague where he was joined by Kepler. It was not until Tycho's death that Kepler could really get to work analyzing the data. It is for his amazing analysis of Tycho's data that Johannes Kepler (1571–1630) is known. It resulted in the three laws of planetary motion, for which Kepler is best remembered. The work is a tour de force of mathematical analysis.

The first law states that the shapes of the planets' orbits are ellipses. An ellipse is like a stretched out circle. An important consequence of this is that the planets are not always at the same distance from the Sun. Kepler's second law states that planets speed up as they get closer to the Sun in their orbits. The third law states that a planet further from the Sun than the Earth will take longer to orbit the Sun, not only because it has farther to go but because it travels more slowly through space. For example, it takes the planet Jupiter about 12 years to orbit the Sun, but Jupiter only has to cover five times as much distance as the Earth does in its orbit. Should Jupiter's year not then equal five earth years? The reason it doesn't is that Jupiter travels around the Sun at a slower speed than the Earth. In many textbooks, ellipses are drawn with a very elongated form. We should bear in mind however, that the orbits of the planets differ from circles at the level of only a few percent. The orbit of mercury, if drawn on this page, would look like a perfect circle to your eyes.

Kepler's work marked a turning point in the history of astronomy. To measure the orbits of the planets as Kepler did is a stunning achievement. Let us remind ourselves that the data he had at hand were accurate planetary positions in the sky. It took quite some ingenuity to calculate the exact shapes of planetary orbits using these data. Using a circular-orbit model for Mars, Kepler predicted that Mars should be off by one quarter of a moon diameter from Tycho's measured position. Kepler noticed that the difference between the predicted position for the planet and the observed position was larger than the errors Tycho quoted for the observed position. He trusted Tycho's error estimates and thus decided that the orbit could not be a circle. In science we always associate error estimates with measurements. I could quote my height as 180 ± 0.5 cm, for example. If a theory predicts my height to be exactly 184 cm, it's not a bad effort, but we can be sure the theory is wrong. The minutiae of error analysis and measurement make possible the quantitative comparison of theory and observation.

Kepler was fascinated by numbers. He thought that the spacings of five solids inscribed within one another could explain the spacings of the planets. Examples of such solids are the cube and the pyramid. The idea was to inscribe an orbit within such a solid so that it just touched the faces. We believe this is a mere coincidence today. Kepler is often described as being something of a mystic, but I find some astonishingly modern ideas in his writing. Speculating about his laws of planetary motion, Kepler thought that the increase in speed of a planet as it gets closer to the Sun might be due to a property of the Sun. He was looking for causal effects. There was something about the Sun that caused a planet to speed up as it approached the Sun. The Sun was somehow acting at a distance and producing an effect on the planets. Kepler thought this force might be magnetic, since a magnet could perceptibly influence the motion of pieces of iron placed at a distance from it. In other words, the Newtonian concepts of force and action at a distance are already present in Kepler's writing.

Kepler argued against Bruno's viewpoint that the universe was infinite. Kepler stated that if the universe was infinite and filled with stars like the Sun the night sky would be much brighter than it actually is. We shall return to this subject later in this chapter.

The Telescope and the Starry Messenger

While Kepler was developing these ideas, laying the foundations for a new cosmology, Galileo (1564–1642) was making surprising discoveries using a new device called a telescope to study the heavens. He saw spots on the Sun, and valleys and mountains on the Moon. These observations were in conflict with the Greek view of the cosmos, which maintained that the heavens were perfect. Galileo also noted that Venus went through a complete set of phases, thereby proving that Venus orbits the Sun and not the Earth. In the Ptolemaic system, Venus is always almost in front of the Sun, so that its maximum phase would be a crescent. Galileo also discovered four moons orbiting Jupiter. This latter discovery clearly showed that moons could orbit a planet while that planet orbits the Sun, thereby removing one chief objection to the motion of the Earth and its moon around the Sun. Galileo also showed that the Milky Way is composed of individual stars. And, in an example of the uneasy marriage between science, technology and society, Galileo showed that the telescope could serve as an early warning system. It could be used to give the inhabitants of a port advance warning of attacking ships, much as another scientific advance, radar, would do in World War II.

Galileo strongly advocated the Sun-centered system and ended up antagonizing the Catholic Church. He wrote a book entitled *Dialogue Concerning the Two Chief World Systems* published in 1632. This book was calculated to make fools of the establishment. The prevailing view of an earth-centered universe was put in the mouth of a naive character called Simplicio. To make matters worse, a third, supposedly unbiased character, continuously sided against Simplicio in the discussions depicted in the book. Galileo probably could have gotten away with much of what he wanted to say if he had done it in a less provocative and polemical manner. Direct provocation of the powers that be is not always a good strategy. Of course, scientifically, Galileo was perfectly correct.

The question still remained as to why the planets obey Kepler's laws. Kepler and, in particular Galileo had developed the beginning of a science of dynamics, but it was Isaac Newton (1642–1727) who would develop a complete theory of motion and gravitation.

Who were the contemporaries of these men of science? Luther, Magellan, Michelangelo, and Leonardo da Vinci were among Copernicus's contemporaries. Galileo, Tycho, and Kepler numbered Shakespeare and Milton as contemporaries while Newton's life overlapped with that of Rembrandt, Voltaire, and J. S. Bach. It is interesting that word of Newton's achievements reached the continent through Voltaire, a French novelist, philosopher and playwright. It is perhaps as if, in more recent times, T. S. Eliot had come back to the United States bringing news of the structure of the atom.

Newton is an intimidating figure in the history of science. Through his mathematical creations, deep physical insights, and inventions he dominated many fields. His theories of motion and gravity broke down the barrier between the heavens and earth once and for all. The same laws that govern the motion of objects on the Earth's surface govern planetary motion. The force that pulls my keys to the ground when I drop them also keeps the Moon in its orbit.

There is something mystical about action at a distance. We say that the force of gravity keeps the Moon in its orbit but no mechanism is specified. Between the Earth and moon lies nothing but empty space, yet the Moon behaves as if a giant string were attached to it, pulling it to the center of the Earth. Described with words, the concept of gravity is not very useful. I might as well tell you that plixes keep the Moon in its orbit. Plixes are, of course, little orange creatures whose job it is to keep the planets in their orbit. The power of Newton's theory lies in the fact that it enables one to calculate the orbits of the planets with very high precision. Newton's theory can be used to plot the trajectories of man made objects in space. Newton's laws also explain the phenomenon of tides. In 1669 Newton was named Lucasian Professor of Mathematics at Trinity College Cambridge, where he remained for many years. Newton published his results in July 1687 in a book entitled *Mathematical Principles of Natural Philosophy*. I was present at the inaugural lecture that Stephen Hawking gave in 1981 when he was elected to that same position. The feeling of a continuous tradition extending hundreds of years is a conspicuous feature of the older European universities.

Newton was undoubtedly a man of genius. Despite his achievements, he does not emerge as a very admirable personality. Maybe this is the price one pays for having the strength of character to do something really outstanding.

Descartes expressed the idea of an evolving unbounded universe. Newton showed that we can discover the fundamental physical laws of matter within the universe. Combine these two ideas and you have the basis of modern cosmology.

The period we are describing was one of world exploration. European nations were sending out ships to look for new lands and riches. These ships had considerable trouble keeping track of their position at sea. North-south position did not present trouble, but east-west position, or longitude, was another story. In 1714 the British government offered a large reward for the solution of this problem. It boiled down to finding a reliable way of telling the time. This could be done mechanically using a clock or using the celestial clockwork. To illustrate this idea, let us imagine that a friend called you up (at midday, while you are having lunch) on his cell phone and said, "I am at sea but I have no idea where I am." You could have a clue as to his location by asking him what time it is. If he says, "Right now, its about the middle of the night," you know that he is on the other side of the Earth from you. He could avoid calling you by having a clock set to your local time. By comparing his local time with yours he would know how far away from you he is. In the absence of cell phones and other modern gadgets, the only way to do this was to find an accurate method of keeping time at sea.

William Hamilton attacked the problem by building increasingly accurate clocks. The astronomers, of course, favored using the relative position of the Sun and the Moon to tell the time. A forerunner of today's global positioning system was proposed: ships anchored at sea would fire cannons at specified times. It is good to be ahead of your time but not too far ahead. The longitude problem led to the creation of the Royal Greenwich Observatory in 1676. It was a rare occasion when astronomy could serve society directly. Realizing in the 1990s that the longitude problem had been solved for quite a while, the British government shutdown the Royal Greenwich Observatory.

Why Is the Sky Dark at Night? Meditations on an Infinite Universe

As early as 1576, Thomas Digges had suggested that the outer sphere in the Copernican model be done away with, that the universe was in fact infinite (Fig. 1.3). This was a natural continuation of the Copernican revolution. Copernicus had removed the Earth from the center of the universe. If the Sun is but one of many

FIG. 1.3 This image first published by Camille Flammarion could be seen as showing man's attempt to look beyond the universe of spheres into the workings of an infinite universe

stars scattered through space, it is plausible that the Sun is not at the center of the universe. Giordano Bruno (1548–1600) even suggested that there are planetary systems surrounding other stars. Almost 400 years later, Michel Mayor and Didier Queloz detected a planet orbiting a star other than our Sun. Currently almost 1,000 planets outside our solar system are known to exist. Bruno also suggested the existence of a universe, without a center, which agrees with our modern ideas of cosmology. Descartes (1596–1650) in his book *Principia philosophiae* proposed the idea that the universe was without center and without limits. He also believed the Sun to be a star like many other stars. He came up with the idea that the universe was not empty but structured with vortices that guide the planets in their orbits. He countered the idea that the universe was created in one moment with the modern idea that the universe evolves with time. This is a key concept of the Big

Bang theory. The question of whether space is full or empty has a long history in cosmology. Newton dismissed Descartes' vortices, but the idea that space was not empty resurfaced with the attempt to understand electric and magnetic fields and the propagation of light through space. This idea was dismissed by experiments and by Einstein in his special theory of relativity. The discovery of dark energy has revived the idea that space is not empty.

When Newton envisioned an infinite universe that was populated by stars scattered randomly in space, he realized that there was a potential problem. If we think of concentric shells surrounding the solar system, we might expect the stars in the more distant shells to exert less pull on the Earth. Gravity is an inverse square law, meaning that the gravitational force between two objects decreases as the square of distance between them. Note, however, that the volume of a shell of finite thickness increases as the square of its radius. In other words, the pull of a star in a shell that is twice as distant as a nearby shell is four times weaker, but the distant shell contains four times more stars. If the universe were slightly more dense in one direction than another, the gravitational forces would be enormous. Newton realized that an infinite universe would have to have a uniform density or the Earth would be subjected to enormous gravitational forces.

Kepler examined the idea of an infinite universe and raised an important objection that was to be argued about for hundreds of years. In 1610 Kepler wrote a short book in which he presented his argument. If the universe is infinite we would expect to see more and more stars as we stare into space in such a way that the night sky should be as bright as the surface of the Sun. It is the same argument as Newton's argument for a homogeneous universe. Both gravity and light flux fall off as inverse-square laws. Although more distant stars are fainter, there are more of them to cover a distant patch of sky. These two effects cancel one another in such a way that we keep adding more and more light until the night sky is as bright as the disk of the Sun. The British astronomer Edmund Halley suggested that the light from distant stars was too faint to have any effect, but this is a mathematically incorrect argument.

The problem of the dark night sky is known as Olbers' paradox. A Swiss astronomer, Jean-Philippe Loys de Cheseaux (1718–1751), suggested in 1744 that some medium lying between the stars was absorbing the starlight. A distinguished German astronomer, Heinrich Olbers (1758–1840), put forth the same hypothesis in 1823. John Herschel (1792–1871), an English astronomer argued that this proposed solution was incorrect because the light from the stars would heat up the intervening medium, which, in turn, would start glowing.

The correct answer to the paradox lies in the finite age of the universe. Light has a finite travel speed, therefore we cannot see the whole universe today but only those regions close enough for light to have reached us since the beginning of the Big Bang. The universe has been around a long time, 15 billion years or so. This is not long enough for the accumulated light from stars to create a very bright sky background. It is interesting that when we take very faint images with large optical telescopes, about a third of the sky is covered with galaxies.

As well as thinking of the universe as a whole, people were turning their thoughts to the structure of the universe. Thomas Wright (1711–1786) published his ideas in a book entitled *An Original Theory of the Universe.* The idea he propounded is that the stars are distributed in a spherical shell around the center of the universe. If we were located somewhere in a sufficiently large shell the stars close to us would appear as if they were distributed in a plane in which we are embedded. This would explain the appearance of the Milky Way in the night sky. Remember that Galileo had shown, using his telescope, that the Milky Way could be resolved into tens of thousands of stars.

Immanuel Kant (1724–1804) suggested in 1755 that the Milky Way is a flattened disk that rotates about its center. This, as we shall see, is our modern view of our galaxy. Kant also believed in the existence of other galaxies–an idea that came to be known as the island universe hypothesis. It is surprising to what extent Kant's cosmological speculations were correct, given the complete lack of supporting evidence at the time.

Pierre Simon Laplace (1749–1827) made a big impact on the public with his book *Exposition du système du monde.* He suggested that the solar system originated from a large flattened slowly rotating cloud of gas and also stated that many other stars may well have their own planetary systems. Laplace also formulated our modern idea of a galaxy containing billions of stars. We then have a universe in which stars are not randomly distributed in space but located inside galaxies. As if this was not enough Laplace came up with the first clear description of what we mean by a black hole, namely an object from which even light cannot escape.

Late Night Thoughts on the Structure of Our Galaxy

William Herschel (1738–1822) took an observational approach to these questions. As well as studying our galaxy, he discovered the planet Uranus and demonstrated that the orbits of double star systems obeyed Kepler's laws. Herschel assumed that stars were uniformly distributed in space. From this it follows that if more stars are counted in a given direction than another it is because the galaxy extends further in space in that direction. Accordingly he set out to map the shape of our galaxy by counting stars in 683 regions of the sky. Using this method of star gauges, Herschel mapped out a shape of our galaxy. Unfortunately, this method is flawed because starlight can be absorbed and scattered by intervening matter, which biases the number counts of stars. One must take into account observational biases in order to draw reliable conclusions from survey data.

Like Galileo, Herschel earned a pension from the government by pleasing the king with one of his discoveries. Herschel had earned his living as a musician before turning to astronomy. He was able to turn full-time to astronomy, thanks to a pension awarded him by King George III. Herschel discovered many nebulae, cloudy patches of light whose origin was uncertain. He managed to resolve some of these into clusters of stars and believed that all nebulae were actually star clusters. In 1790, Herschel discovered a planetary nebula, an object consisting of a cloud of gas surrounding an evolved star. This confused the issue for Herschel, since the nebula itself clearly could not consist of stars. One hundred years after his observation the first photographs of planetary nebulae were taken. Roughly 100 years after that, the first photographs of planetary nebulae were taken with the Hubble Space Telescope. The improvement in observing techniques is really impressive, as the contrast between photographs taken 100 years apart can testify.

With the work of Herschel we see for the first time statistical methods in action and also the planning and execution of a major survey. This approach has proven to be the most fruitful one in modern cosmology. It is not very glamorous, because it is laborious and repetitive work that can take several years, but, in the long run, surveys yield the most information. Herschel made a catalog of 2,500 nebulae, which included the famous list of 103 objects compiled by the French astronomer Charles Messier (1730–1817). He found that 29 of the Messier nebulae were in fact collections of stars. Messier's catalog had originally been compiled to help comet hunters to distinguish the fuzzy nebulae from comets–that is, to separate objects at the time considered boring from objects of genuine interest. Many of the "boring" objects in Messier's catalog are currently the subject of very detailed studies at all wavelengths.

Herschel carried out his work using a reflecting telescope with a 47 cm mirror, although he earned fame and fortune by discovering Uranus with a telescope with a 15 cm diameter mirror. The method of star counts was also used by the Dutch astronomer Jacobus Kapteyn (1851–1922) to build a model of the galaxy. Kapteyn searched for evidence of interstellar absorption but found none. According to Kapteyn's model, the galaxy was about 45,000 light years in diameter, with the Sun lying about 2,100 light years from the center. Kapteyn's work involved collaboration between astronomers in many countries. It is one of the very positive aspects of the scientific enterprise that people from different countries and different cultural traditions can work together to achieve common goals.

The Realm of the Nebulae

Simon Marius (1573–1624) published the first observations of an island universe made with a telescope. He reports observations made on 15 December 1612 of what we now know is the Andromeda galaxy, a large spiral galaxy similar to our milky way galaxy. Marius was unaware of the nature of Andromeda but he gives an accurate description of its appearance seen through a small telescope. About 100 years later, an anonymous author believed to be Halley published "An Account of several Nebulae or lucid Spots like Clouds, lately discovered among the Fixt Stars by help of a

Telescope". The fact that these nebulae remained fixed relative to the stars suggested they were very far away from the solar system.

In 1845, William Parsons, the Earl of Rosse, built a reflecting telescope with a 72-inch mirror. With this instrument he established that some of the nebulae have spiral structure. The nature of these nebulae was by no means clear. Were they spiral-shaped clouds of gas inside our own galaxy, or were they huge galaxies in their own right, appearing small and blurry only because they were so far away?

Clues to resolving this problem were provided by William Huggins (1824–1910). Huggins was a pioneer of spectroscopy, a technique astronomers use to analyze the detailed color of the light emitted by stars and gas clouds. A spectrum consists of the brightness of the light detected at different colors. The elements of the periodic table leave their imprint on the light in the form of excess absorption or emission of light at certain very specific colors or wavelengths. Huggins and Miller in 1864 showed that there was a similarity between spectra of stars and the spectrum of the Sun which Fraunhofer had published in 1814. Huggins next took spectra of the nebulae that Herschel called planetary nebulae. The spectra he obtained were quite different to stellar spectra. The planetary nebulae spectra contained strong emission at certain specific wavelengths. It took until 1927 to realize that the emission was coming from twice ionized oxygen. It was clear to Huggins that the emission from planetary nebulae was due to gas and not distant star clusters. When Huggins took a spectrum of M31, the Andromeda nebula, he found that its spectrum resembled that of star clusters. The result implied that maybe this nebula was a distant collection of stars. In fact Huggins had obtained several spectra of nebulae that were different to planetary nebulae in that they did not show strong emission line of oxygen and hydrogen. This work implied that the spiral nebulae discovered by the Earl of Rosse were indeed distant aggregations of stars.

The issue of the distance to the nebulae was finally resolved by Edwin Hubble (1889–1953), who on the 6th of October 1923, discovered a variable star in the Andromeda nebula. Allan Sandage (1926–2010), Hubble's student, who was born 3 years after that discovery, kept the photographic plate on which Hubble wrote "VAR!" in red ink next to the variable star he discovered. The plate is shown in Fig. 1.4. The luminosity of a variable star can

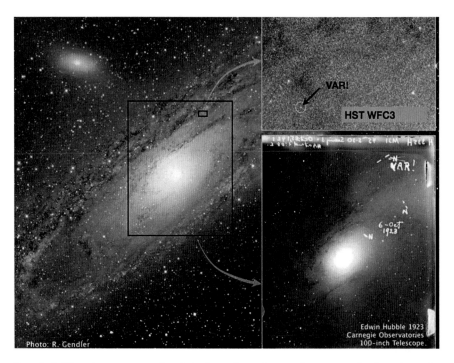

FIG. 1.4 The image on the *left* is taken from the ground showing the Andromeda galaxy. The image at the *bottom right* is Edwin Hubble's discovery image of a cepheid in the Andromeda Nebula. These variable stars were crucial to determining the distance to the Andromeda Nebula and showing that it was not part of the Milky Way galaxy. This same cepheid star was observed many years later by the Hubble Space Telescope (see image at *top right*). The boxes in the *left hand* image show the location of the two images on the *right hand* within the galaxy (Credit: E. Hubble, NASA, ESA, R. Gendler, Z. Levay and the Hubble Heritage Team)

be inferred from its period. It is as if the wattage of a light bulb in a light-house could be inferred from how fast the light appears to turn on and off. These stars can therefore be used to measure distances.

The most useful variable stars for measuring the distances to galaxies are called Cepheid variable stars. One can determine the luminosity of a Cepheid star to an accuracy of about 10 % by measuring the time period over which its brightness varies. The Cepheid star that Hubble discovered was clearly in the Andromeda nebula and was at such a distance that it could not be in our galaxy thus proving that the Andromeda nebula lay outside our own galaxy. By the end of 1924, Hubble had discovered 36 variable stars in Andromeda, 12 of which were Cepheid variables. Using those stars he concluded that Andromeda was 900,000 light years away. Today we believe Andromeda to be about 2 million light years away.

In the roughly 70 years between the discovery of spiral nebulae and Hubble's solution of the problem, there was much discussion about their nature, which culminated in a famous so-called debate conducted in April 1920 at the annual meeting of the National Academy of Sciences held at the Smithsonian Institution. On one side of the debate, was Harlow Shapley (1885–1972) who argued that the Milky Way was so large that the spiral nebulae had to be within it. Shapley believed the Milky Way to have diameter of 300,000 light years. Heber D. Curtis (1872–1942) disagreed, holding that the nebulae were "inconceivably distant galaxies of stars, or separate stellar universes, so remote that an entire galaxy becomes but an unresolved haze of light". Curtis thought that the Milky Way's diameter is 30,000 light years. The debate attempted to settle three key questions:

1. What are the distances to spiral nebulae?
2. Are spiral nebulae composed of stars or gas?
3. Why do spiral nebulae avoid the plane of the Milky Way?

The key argument concerning the distances to spirals centered on some measurements made by Adriaan van Maanen (1884–1946) a Dutch astronomer. Van Maanen attempted to measure proper motions of spiral galaxies. By proper motions we simply mean the motion of objects in the sky. Proper motions are caused by the motion of the object itself, as opposed to parallax, which, as we have seen, is an effect caused by the Earth's motion around the Sun. He compared photographs of a given spiral taken some time apart and detected motion of the spiral nebulae. If Shapley's model of the Milky Way was correct and other spiral nebulae were of the same size as the Milky Way, van Mannen's measurement implied that they were rotating close to or faster than the speed of light. To avoid this conclusion, Shapley decided that the spiral nebulae were smaller objects within our own galaxy.

Curtis decided to ignore van Maanen's data, believing them to be spurious. Curtis was correct: The proper motion measurements were in error. When one looks at the data published by van

Maanen in the *Astrophysical Journal*, they seem highly convincing. It is hard to imagine what observational error could produce the effect that he measured. On reading one of van Maanen's papers, I noticed, interestingly, that one of the photographic plates he examined was taken by Curtis. It is probable that, as an observer, Curtis had a good feel for the limitations of the data. To measure the positions of blobs of emission in spiral galaxies to an accuracy of 1/30,000 of a degree using photographic plates taken at the turn of the twentieth century sounds like an impossible task. Yet van Maanen claimed to have done precisely this. One moral to be drawn from the so-called Great Debate is that the two protagonists drew more reliable conclusions when discussing data that they were fully familiar with.

As to whether spirals were composed of gas or stars, Shapley argued in favor of the former. Spiral nebulae appeared bluer in their outer portions than in their centers, suggesting they did not consist of stars. Those favoring the view of spirals as galaxies argued that we could not resolve the individual stars because the galaxies were so far away.

Who, If Anybody Won the Great Debate?

Why do spiral nebulae avoid the plane of the Milky Way? Shapley argued that since they avoid the Milky Way plane they must be influenced by it and therefore be close by. Curtis countered by saying that many spiral nebulae exhibit a central belt of obscuring material. If we imagine that the Milky Way also has such a belt and that spirals are external to the Milky Way system, the zone of avoidance could be explained. The spiral galaxies are blocked from our view by the dust, but, in fact, are really actually there. If we imagine the Milky Way as a disk containing dust, then, when we look in the plane of the Milky Way, the dust blocks our view of the galaxies, whereas when we look perpendicular to the plane of our galaxy, we see many galaxies because we do not have to look through as much dust. The Sun looks red at sunset for similar reasons. When the Sun is on the horizon at sunset, the sunlight passes through more of the Earth's atmosphere than at noon. The atmosphere contains molecules that scatter the blue light from the Sun, making the Sun look redder. Again, Curtis' explanation was correct.

When looking at the protagonists and the arguments used in the debate we see how confusing science can be. Neither Shapley nor Curtis was 100 % correct. Some key measurements that had been recently published by a reputable scientist (a former student of Kapteyn) proved to be incorrect. Both Shapley and Curtis made reasonable arguments to support their cases. We currently believe the Milky Way to be three times larger than Curtis thought and three times smaller than Shapley believed it to be. To summarize, one might say that Shapley made some good arguments, and came up with the wrong answer; Curtis did not necessarily have the best arguments but got the answer right. For want of a better word, intuition can play a big part in arriving at one's conclusions when the evidence is incomplete or misleading.

The opinions of both Shapley and Curtis were most reliable when they were discussing observations they had obtained themselves or were familiar with. The moral for astronomy students is to "know the data". When one is familiar with observations, one has a feel for the experimental accuracy and possible errors that is hard to get by simply reading the literature.

A key issue in the Great Debate was that of the amount of absorption of starlight by gas and dust. Robert Trumpler (1886–1956) was the first to demonstrate the existence of an absorbing medium in-between the stars. Born in Switzerland, Trumpler came to work in the United States, spending most of his research career at Lick Observatory, where he studied star clusters. He could measure their angular size on the sky (by 'sky' in this case we mean 'on a photographic plate') and used this to estimate their distances, assuming these clusters had similar physical sizes. He found that the more distant clusters appeared to be intrinsically less luminous than the closer ones. He chose not to take this fact at face value but postulated instead the existence of an absorbing medium, which would make distant objects look fainter than they really were. This could immediately explain why the external spiral nebulae avoided the plane of the Milky Way. Spiral galaxies behind the plane of our galaxy are obscured by it. More direct evidence for an interstellar medium came from spectra of binary stars. These stars are orbiting each other and produce spectral lines that shift back and forth in wavelength, reflecting the motion of the stars radially toward and away from the observer. These spectra contained some lines that did not shift at all, suggesting that they were associated not with the stars, but with an interstellar medium. Taking into account interstellar absorption helped explain why astronomers were calculating different sizes for the Milky Way using different methods.

During the late 1920s people were analyzing the motions of stars in our galaxy and concluding that our galaxy must consist of a disk that rotates and a halo that does not. Most of the key insights were provided by Jan Oort (1900–1992). Oort began his studies at Leiden and went on to study in Groningen under Kapteyn. Like Kapteyn, Oort promoted international collaboration and was instrumental in bringing about the creation of the European Southern Observatory. He left his substantial mark on many fields of astronomy, from the study of comets up to cosmology. Oort showed that our galaxy rotates differentially, that is to say not as a solid body. From the study of stellar motions perpendicular to the galactic plane and close to the Sun, Oort showed that, in the solar neighborhood of the galactic disk, twice as much mass must exist gravitationally as can be accounted for by luminous matter. This is known as the missing mass problem–something of a misnomer, since we believe the mass is actually there. The problem is not that the mass is missing but that it does not emit light. The mass is present but invisible because it does not emit light. The missing mass problem is one that pervades all of cosmology, and we shall encounter it again and again. To infer the presence of dark matter one must make use of a theory of gravitation, namely Newton's theory or its more accurate extension developed by Albert Einstein, General Relativity. This theory has not been tested on the physical scales (thousands to millions of light years) on which we are applying it. There is the possibility that gravity does not follow an inverse square law on these large scales. Nevertheless, most astronomers prefer to admit the existence of dark matter.

The story of the discovery of the structure of our galaxy and the existence of external galaxies is not as elegant as that of planetary motion. This is because a number of key discoveries whose interpretations influenced each other were made in a short span of time. It may also be that we have a more accurate picture of relatively recent events that are better documented than debates taking place several 100 years ago. It is certainly fascinating to look at the actual data that were published in the 1920s and 1930s. As happens today, some of the observational evidence was rather thin at times.

Percival Lowell (1855–1916) founded the observatory that is named after him in 1894. Lowell was fascinated by the idea that there might be life on Mars. In 1901, he hired a man named Vesto Slipher to assist him in his work. Lowell also was interested in the spiral nebulae, because he believed they were possibly solar systems in formation. This hypothesis implied they should rotate, and Slipher used a spectrograph to measure the shift in the wavelength of spectral lines emitted by atoms in gaseous form and hence infer the speed at which these atoms are moving. The wavelengths of spectral lines can be measured in the laboratory and compared with the wavelengths of lines measured with a spectrograph. In 1912 Slipher had obtained a spectrum of the Andromeda nebula and concluded that it is approaching the Earth with a velocity of 300 km per second. This was a surprisingly large speed. The Andromeda nebula is unusual in that it is the only large galaxy that is approaching the Milky Way galaxy. We estimate that Andromeda will collide with the Milky Way in about 4 billion years.

By 1914 Slipher had obtained spectral line shifts for 14 spiral nebulae. These shifts were all toward the red and are commonly known as redshifts. The fact that shifts were towards the red and quite large implied that they had large positive radial velocities. In other words, the spiral nebulae appeared to be moving away from the Milky Way at high speeds. Interestingly, the high recession speeds of the spirals were used as an argument against the island universe hypothesis. Slipher also measured rotational velocities for the spiral nebulae. His results contradicted the proper motion results of van Maanen! Both thought the spirals were rotating, but they disagreed about the rotation direction. We shall discuss these rotational velocities in more detail when we discuss the nature of dark matter in the universe. Slipher achieved worldwide fame for his spectrographic observations of the nebulae and eventually became director of Lowell Observatory. It was under his directorship that the planet Pluto was discovered in 1930 by Clyde Tombaugh.

The Expansion of the Universe

The situation, even after the Great Debate, was one of great confusion. It was by no means clear if the spiral nebulae lay inside or outside the Milky Way. Hubble finally solved the problem by measuring distances to variable stars in spiral nebulae. By 1929 Hubble had obtained distances for 18 of the 46 objects for which velocities were then available. The most distant of Hubble's nebulae was moving away from us at 1,000 km per second. Hubble estimated its distance as 700,000 light years, thereby placing this nebula well beyond the edge of the Milky Way, even according to Shapley's large diameter for our galaxy of 300,000 light years.

The result that was to make Hubble famous was the correlation he found between distance and velocity. He found that the more distant nebulae were moving away from us faster than the nearby ones. The proportionality constant between these two quantities is now known as Hubble's constant. Hubble was unaware that there are two kinds of variable stars. Because of this, he underestimated the distances to the nebulae by a factor of ten or so.

Hubble does not deserve all the credit for the discovery of an expanding universe. In 1927 George Lemaître (1894–1966) published his model of an expanding universe (Fig. 1.5). He predicted a relationship between velocity and distance in the form that we now credit to Hubble. He went further and used data to calculate a value of the Hubble constant (roughly 8 times larger than the accepted value today). Hubble found the same relation 2 years later using data that had mostly been gathered by Slipher. It is possible that Lemaître's paper did not have a great impact because it was published in French. Even more intriguing is the fact that the version of the paper which was published in English with the help of Arthur Eddington in 1931 does not contain the estimates of the Hubble constant from the 1927 paper.

One can use Hubble's constant to get an estimate of the age of the universe. Hubble's estimate of his constant was about ten times too big, implying an age of the Universe of 2 billion years–rather discomforting, given that the oldest rocks on Earth were believed to be about 4 billion years old. Conflicts with geologists had already arisen when physicists in the nineteenth century estimated the age of the Sun to be 100 million years,

FIG. 1.5 Georges Lemaître who proposed the idea that later became known as the Big Bang theory of the origin of the universe (Credit: Archives Georges Lemaître, Université catholique de Louvain, Centre de recherche sur la Terre et le Climat G. Lemaître, Louvain-la-Neuve, Belgique)

which contradicted the evidence from geology. When Hubble's distance scale was revised, it resulted in a universe that was at least 10 billion years old, the most recent estimate being about 14 billion years.

Hubble also studied the shapes of the nebulae on photographic plates and produced a classification scheme that is still used today. Not all nebulae have a spiral shape. Some appear elliptical. Hubble proposed a scheme suggesting that elliptical nebulae might have evolved into spiral nebulae.

While the debates about the nature of our galaxy and the spiral nebulae were fermenting, Albert Einstein (1879–1955) was working on including the concept of curved space into a theory of gravitation. This work was done in the years 1912–1914. The theory produced a set of equations that could be used to express the geometry of the universe. The simplest solution he found to these equations implied that the universe was expanding. By this we mean that the distances between objects in the universe are increasing. This conclusion in 1917, came about 10 years before the formulation of Hubble's law. Since Einstein believed that we live in a static universe, he modified the equations to produce a static universe proposing the constant Λ. Thus Einstein very nearly predicted–but not quite–the expansion of the universe.

Interestingly, de Sitter (1872–1935) had stated in 1917 that one of the solutions of Einstein's equations indicated a displacement of spectral lines for distant objects. This effect, now known as the redshift, is just what Hubble was to observe in the decade that followed. Willem de Sitter was appointed Professor of Astronomy at Leiden in 1908. His portrait hangs in the hall of Leiden observatory where astronomers (at least in my time) gathered for their morning coffee. I used to have an occasional beer in a bar in Leiden in the 1980s. I remember being puzzled when I saw a small portrait of Einstein almost hidden away in a corner. I asked the owner why this portrait was there. He said that Einstein had frequented this bar when he had visited Leiden. He must have come there to relax when he visited de Sitter.

The Idea of the Big Bang

The expansion of the universe implies that the universe has evolved from a dense state. This theory or idea is now known as the Big Bang. In the 1930s it was realized that nuclear reactions provided the energy to fuel the stars. Might it not be possible that nuclear reactions had taken place in the early universe when temperatures reached millions of degrees? In 1948, George Gamow (1904–1968), a Russian physicist working in the United States, considered this hypothesis. He predicted that if a hot Big Bang had indeed taken place, some radiation left over from the intense heat of the early universe should be present today. It was not until 1965 that this radiation was discovered by scientists who were not trying to test this theory at all. They were scanning the sky in search of background radiations that interfered with communications satellites.

We have come a long way from the start of this chapter, which began with the Hopi creation myths. We went on to discuss the Greek concept of the universe and focused on the problem of explaining planetary motion. This was the central cosmological problem for almost 2,000 years. The universe consisted of a few thousand stars and the planets wandering among them.

The attempt to solve this problem led to the formulation of a theory of motion and gravity culminating in the work of Newton. We discussed problem of the dark night sky (Olbers' paradox) as viewed by various thinkers. The next major historical theme concerned the nature of our galaxy and its relation to other galaxies. The realization that the spiral nebulae were large galactic systems led to the concept of an expanding universe.

What drives the growth of knowledge? Tools provided by developments in technology are essential to this process. The discovery that planets move on elliptical orbits was dependent on the measurements made with Tycho's precision instruments. Galileo's telescope increased light gathering power and the amount of detail that could be seen (resolving power) by a factor of ten. This enabled him to remove any doubts about the Sun centered model of the solar system. The same is true in the field of theoretical astronomy. Many of the ideas that Newton used in his theory of motion and gravitation were not original. Newton did however invent a crucial mathematical tool (calculus) that made it possible to explain the elliptical orbits of the planets as well as the motion of comets and projectiles on earth.

Martin Harwit has put it best in an article published in the journal Physics Today:

> The history of science so often alludes to the importance of great ideas. That notion needs to be carefully qualified. In astrophysics, new ideas are afloat all the time. Ideas are, of course, needed. But at critical junctures in the history of astronomy, there is generally an overabundance of ideas on how to move ahead. Supporters of the various ideas debate them vigorously, mostly with no clear-cut outcome. Resolution is usually attained only with the arrival of new tools that can cut through to new understanding and set a stagnating field in motion again.

In the next chapter we discuss the Big Bang theory. The concept of expansion suggests that the universe was different in the past. We have run the clock forward from about 300 BC to the mid twentieth century in this chapter. In what follows, I will take you back in time to a split second after the Big Bang. It is quite amazing that we can discuss in a sober manner what happened in the universe a few minutes after its creation.

Further Reading

Man Discovers the Galaxies. R. Berendzen, R. Hart and D. Seeley. New York: Science History Publications, 1976.

Discovering the Expanding Universe. H. Nussbaumer and L. Bieri. Cambridge: Cambridge University Press, 2009.

The Book Nobody Read: Chasing the revolutions of Nicolaus Copernicus. Owen Gingerich. New York: Walker and Company, 2004.

Cosmic Discovery; The search, scope, and heritage of astronomy. Martin Harwit. New York: Basic Books, Inc., 1981.

The Discoverers. Daniel Boorstin. New York: Random House, Inc., 1983.

Longitude. Dava Sobel. New York. Walker and Company., 1995

The Day we Found the Universe. Marcia Bartusiak. New York: Random house, Inc., 2009.

The Cosmic Century, A History of Astrophysics and Cosmology. Malcolm Longair. Cambridge University Press, 2006.

2. The Three Pillars of the Big Bang Theory

The evolution of the Universe can be compared to a display of fireworks that has just ended: some few wisps, ashes and smoke. Standing on a cooled cinder, we see the slow fading of the suns, and we try to recall the vanished brilliance of the origin of the worlds.

Abbé George-Henri Lemaître

Evidence for the Big Bang

The Big Bang theory has gained general acceptance by its ability to explain in a simple manner three key cosmological observations. These three observations are (1) the expansion of the universe as measured by the redshift of light emitted from galaxies (2) the existence of the cosmic background radiation and (3) the relative amounts of hydrogen, helium and deuterium in the universe. The theory states that the expansion of the universe began at a finite time in the past, in a state of enormous density and pressure. As the universe grew older it cooled and various physical processes came into play which produced the complex world of stars and galaxies we see around us.

The Big Bang theory enables us to understand many different facts about the universe in a cohesive manner. Almost all astronomers believe it to be the best theory of the universe. Let us keep in mind, however, that the finest scholars were completely mistaken about the nature of our universe for most of recorded history. We present-day scientists can only do what any jury does: find the theory that is most consistent with the available facts. We now turn to the details of the three observational pillars of evidence for the Big Bang theory.

G. Rhee, *Cosmic Dawn: The Search for the First Stars and Galaxies,*
Astronomers' Universe, DOI 10.1007/978-1-4614-7813-3_2,
© Springer Science+Business Media, LLC 2013

The First Pillar: Hubble's Law

In Chap. 1, we learned of the history of astronomy culminating in the debate concerning the nature of the spiral nebulae. The correct answer turned out to be that the spiral nebulae are galaxies much like our own. Vesto Slipher had determined that these galaxies are moving away from us at very high speeds. By measuring their distances, Edwin Hubble demonstrated that the galaxies farthest from us are moving away at the highest speeds–Hubble's law. Let us see what this implies.

Imagine a race among three cars. Car number one travels at a steady speed of 60 miles an hour, car number two travels at a steady speed of 70 miles an hour, and car number three travels at a steady speed of 80 miles an hour. One hour after the start of the race we check on the location of the cars. Car one has gone 60 miles, car two 70 miles, and car three 80 miles. Seen from the starting gate, these cars obey Hubble's law. The car that is farthest from us is the one that is traveling fastest. The first and third car are now separated by a distance of 20 miles after 1 h. In another hour, their separation will have grown to 40 miles. By looking at the cars 1 h after the race started and working back in time, we can conclude that they all left the starting line at the same time. Since galaxies obey Hubble's law, we can draw the same conclusion. In the past, the galaxies were all closer together, and, in the future, they will be farther apart. We thus encounter one fundamental property of the universe: it is evolving. The universe looked different in the past and will look different in the future. The reader may ask why the cars acquire different initial speeds. This is where our simple analogy breaks down. When we look in more detail, we see that, in fact, it is the fabric of space that is expanding rather than objects moving in space.

A simple estimate of the age of the universe can be obtained using Hubble's constant. We can calculate for a given galaxy, knowing its current distance and velocity how long it took to reach its distance from us. This time is equal to (d/v) which turns out to be one divided by Hubble's constant for every galaxy. This rough estimate gives an answer of 14 billion years for the age of the universe, quite close to our correct answer of 13.73 billion years calculated using the proper cosmological models.

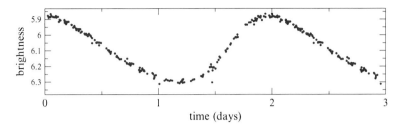

FIG. 2.1 The variation of brightness with time for the Cepheid variable star SU Cas measured by the Hipparcos satellite. The time interval from peak brightness to the next peak is about 2 days. We can use the pulsation time to infer the total luminosity of this star and hence its distance from us. This star is about 1,500 times as luminous as the Sun and 6 times as massive. The distance to the star is about 1,400 light years

Measurements of Redshift and Distance

Hubble's law can be stated in the equation $v = H \times d$. Establishing Hubble's law involves measuring distances (d) and velocities (v) of galaxies. In practice, distance is much harder to measure than velocity. We measure the distances to galaxies in a number of ways, but the principle involved is always the same. We have to be able to identify objects of known luminosity in galaxies. We know how luminous the Sun is. If, for example, we could identify a star like the Sun in another galaxy we could then note how much fainter than the Sun that star appears to us. If the star was a million times fainter than the Sun in the sky, that star would be a 1,000 times farther from us than the Sun is. The problem of measuring distances in astronomy boils down to finding reliable indicators of distance. The indicators used by Hubble were variable stars. Figure 1.4 shows the image that Hubble used to discover a variable star in the Andromeda galaxy. These stars are still used for that purpose today. They vary in brightness because they pulsate (Fig. 2.1). It turns out that pulsation time depends on luminosity. A star with a slow pulsation time is brighter on average than a star with a short pulsation time. A star with a 3-day pulsation period is about one 1,000 times more luminous than the Sun, whereas a star with a 50 day pulsation period is about 10,000 times as luminous as the Sun. By knowing the true luminosity of a star and comparing it with the star's apparent brightness, we can calculate the distance to that star and hence to the galaxy containing that star.

But how do we determine the true luminosities of stars? How was the original period luminosity relation discovered? We can determine the distances to nearby stars using parallax and hence deduce their luminosities. Parallax is the motion in the night sky of nearby stars relative to distant stars over a period of months. Parallax is caused by the Earth's motion around the Sun, it can be used to estimate distances to nearby stars once the Earth-Sun distance is known. We see that variable stars whose distances are known exhibit a period luminosity relation, and we use this relation to deduce the distances to more distant variable stars.

Distance estimation in astronomy is difficult. Absorption of starlight by dust and gas can make stars look dimmer than they really are. Also, nature has contrived to make different kinds of variable stars, and we must not get them confused. Hubble thought that only one kind of variable star existed. This error caused him to underestimate the distances to galaxies by a factor of 10. Today we measure the distances to nearby galaxies to an accuracy of a few percent. For example if we say that a galaxy is 10 million light-years away, a 10 % error means that its distance may lie anywhere between 9 and 11 light years.

What about galaxy velocities? To understand how we measure velocities we must understand something about light and atoms. Atoms are built from particles called electrons, protons, and neutrons. The protons and neutrons are located in the nucleus of the atom, and the electrons orbit this nucleus. Atoms consist mostly of empty space. If you make a fist and imagine your fist is the size of an atomic nucleus, then the atom is as big as the US Capitol and if it happens to be a hydrogen atom then it has a single electron like a moth flitting about in an empty cathedral.

Electrons in atoms of a given element can only have certain specified energies. The electrons can change their energy by emitting and absorbing light or by colliding with other atoms. An atom of a given element, say hydrogen, can emit light only at specific energies or wavelengths. The wavelength of light is a detailed measure of its color. Broadly speaking, red light has a long wavelength and blue light has a short wavelength. Each element (hydrogen, helium, lithium, and so on) has its own set of wavelengths

FIG. 2.2 We measure redshifts using the patterns that are present in the spectra of galaxies. The *upper* spectrum shows schematically where the dark narrow absorption bands might appear in a spectrum of the Sun (in practice we see many more features than are shown). These features would be shifted towards the *red* for a galaxy as shown in the *lower* spectrum. The amount of the shift tells us how fast the galaxy is moving away from us, about 7 % of the speed of light in this case. We can then use Hubble's law to compute the distance to the galaxy in question, which turns out to be about a billion light years, in this example

associated with it, much like a fingerprint. Figure 2.2 illustrates that atoms can absorb light in a very narrow band of wavelengths and leave their mark on the broad colors emitted by stars and galaxies. It is one of the triumphs of twentieth-century physics that we can calculate these wavelengths theoretically. These developments started with the work of Einstein, Bohr and their students. The work culminated in the development of quantum electrodynamics by Feynman, Dyson, Schwinger and Tomonaga.

There is one more detail you need to know. When light is emitted by an atom that is moving away from us, its wavelength is shifted toward the red end of the spectrum. When the atom is moving toward us, the light is shifted toward the blue end of the spectrum. Sound waves behave in a similar manner. We can all hear how the pitch of a car engine increases when the car is moving toward us and decreases when the car is moving away. This effect is known as a Doppler shift. The effect is common to sound and light because both these disturbances propagate as waves. In the case of light, the effect is known as the redshift. Slipher and Hubble observed that the light coming to us from galaxies was shifted toward the red and inferred that these galaxies are moving away from us. You may think we have misidentified the light from the galaxies and that what we think is hydrogen at high redshift is say neon at zero redshift. The answer is that the light we receive from galaxies consists of a pattern of emitted and absorbed light known as a spectrum. From one galaxy to the next we see the whole pattern shifted towards the red and it is hard to make an error. The concept is illustrated in Fig. 2.2.

Hubble's observations implied that, in the past, galaxies were closer together. In other words, when the universe was younger, it had a higher mean density. The universe began in a state of very high density and has been expanding and decreasing in density ever since. This is the idea at the heart of the Big Bang theory.

How We Use Redshifted Light to Look Back in Time

Imagine the universe is a loaf of bread with raisins embedded in it. If the loaf of bread expands, the raisins move farther apart from each other. We measure how much the loaf (space) has expanded using a quantity called the expansion factor. If the expansion factor doubles over a given time interval, then the distance between any two raisins (galaxies) has also doubled. The concept of expansion factor gives us another way of thinking about the redshift.

Let us imagine that at a certain time light is emitted from a distant galaxy with a wavelength λ_1. We receive the light in our telescope at a later time when the light has a longer wavelength λ_2. We calculate the redshift from the ratio of the two wavelengths λ_2 and λ_1. The ratio of the two wavelengths is just the amount by which the universe has expanded during the time the light traveled from the distant galaxy to Earth. If the universe expanded by a factor 2, the wavelengths will have stretched by a factor 2. We believe the universe to have been expanding since the Big Bang, so the expansion factor has always increased. Our current belief is that the expansion will go on forever. In fact, in a few billion years astronomers in our own milky way galaxy (which will have merged with the neighboring Andromeda galaxy) will not even be able to see any galaxies at all with their telescopes.

There is a simple relation between expansion factor and redshift. The expansion factor is equal to one plus the redshift. If we observe a galaxy at a redshift of 5, we can say that the universe has expanded by a factor 6 in the time since the light from that galaxy started on its way to us. This provides us with another way

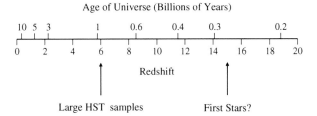

Age of Universe (Billions of Years)

FIG. 2.3 A timeline of the universe comparing the age of the universe with redshift. The redshift measures how much the light emitted by a distant galaxy has been shifted towards the red due to the expansion of the universe. Galaxies which emit light at very early times have very high redshifts. The redshift is an observable quantity. Using known values for the Hubble constant and the density of the universe we can reliably estimate the age of the universe at the time the redshifted light was emitted. The goal of observational cosmology is to explore with observations the area between redshift 6 and redshift 14. We want to know how and when the first stars and galaxies formed in the first billion years after the Big Bang (Credit: Rychard Bouwens)

of thinking of the redshift. Rather than think of galaxies moving away at tremendous speeds, we can think of the wavelength of the emitted light increasing as the light travels on its journey to us. Since we are not at the center of the universe, one might think that certain galaxies should have blueshifts as they move away from the center toward us. In fact, if we consider a string of equally spaced galaxies in a line that appear to obey Hubble's law as seen from one, then an observer on any of these galaxies will think that he is at the center of the universe. This is an application of the cosmological principle, which states that the universe, on average, looks the same to any observer located within it.

The redshift is an observable quantity, which we can calculate using a galaxy spectrum obtained at a telescope (see Fig. 2.2). Using our models we can compute the age of the universe at the time redshifted light set off on its journey to us. Figure 2.3 illustrates the relation between the redshift and the age of the universe at the time the light was emitted. As the redshifts increase we can look back over aeons of time to a period when the universe was a small fraction of its present age. With our telescopes we can see galaxies out to redshifts of about 8 corresponding to a time when the universe was about 600 million years old, about 5 % of its present age. The Earth is about 4.6 billion years old, and the oldest rocks on Earth are about 2 billion years old. These rocks, therefore, tell us what conditions were like on our planet when it was slightly over half of its present age. We astronomers, however, can look back in time and see objects as they looked long before the Earth or the Sun had even formed. There is a price we have to pay for this amazing view of the universe. The light of stars like our Sun that peaks in the visible part of the spectrum gets shifted in wavelength into the infrared part of the spectrum for high redshifts. To explore the high redshift universe we need infrared detectors on the largest ground based telescope and also in space.

When we look at the Moon we see it as it was 1 s ago, because that is how long light takes to reach us from the Moon. When we look at the Sun, we see it as it was 8 min ago, because that is how long it takes for light to travel the distance from the Sun to the Earth. The nearest star is 4 light years away. In the movie *Contact*, Jodie Foster used a radio telescope to detect a signal from an alien civilization. The aliens had received our first TV broadcast made in the 1930s and beamed it back to us. Since we received the broadcast 60 years after it was first beamed into space, and assuming the aliens beamed it right back, we could conclude that the aliens were located 30 light years away from us. Imagine a phone conversation where you had to wait 60 years to get a reply! The light travel time to the large spiral galaxy nearest to us, the Andromeda nebula, is 2 million years. You can see this object with your naked eye on a dark night. Keep in mind that the light that hits your eye is 2 million years old. The light from the most distant galaxies that we can see was emitted so long ago that these objects would look quite different if we could see them as they are today.

Astronomy has something in common with geology. Both disciplines deal with an "experiment" that has been run once. Both study the past over immense expanses of time. Geologists do this by studying rocks and fossils, astronomers use telescopes to actually look into the past. Just as geologists see that species in the past were different from the species we see around us today, astronomers see that very distant galaxies look different from nearby galaxies. The distant galaxies are seen as they were when they were young. Indeed, our journey back in time through the universe is like a journey through the strata of the Earth.

The Second Pillar of the Big Bang:
The Cosmic Background Radiation

In 1967, the discovery of the cosmic background radiation provided strong support for the Big Bang theory. In fact the radiation's existence had been predicted 20 years earlier. To understand the importance of the background radiation we have to first consider what happens to matter in the early universe.

The density of the universe was higher in the past than it is today since objects were closer together. But just what do we mean by the density of the universe? To estimate the density of the universe, we take a large volume of space–say a cube 100 million light years on a side–and estimate how much mass lies in this volume. The density is obtained by dividing the mass by the volume. If we add up the amount of mass in stars and the gas between the galaxies, we arrive at an average density of 0.1 atoms per cubic meter. When we go back to a redshift of five (the redshift of distant galaxies), the expansion factor has shrunk by a factor six, so the volume of the cube will have shrunk by a factor 6^3 or ~ 200. The density of the universe would thus be 20 atoms per cubic meter at that time. When we look back at a galaxy of redshift five, we know that the density of the universe at the time that light was emitted was about 200 times the density of the universe today.

Atoms such as hydrogen and helium only account for a few percent of the total observed density of the universe. Most of the mass in the universe is in a form other than atomic matter. Dark matter plays such an important role in cosmology that it will be the subject of Chap. 4. We shall also encounter a mysterious substance called dark energy. Brian Schmidt, Saul Perlmutter and Adam Riess, the leaders of the two teams that proved the existence of dark energy were awarded the 2011 Nobel prize for physics.

The origin of the cosmic background radiation lies in the interaction of light and matter. Metal objects that are heated to a sufficiently high temperature start to glow. Radiation emitted by objects of known temperature is called black body radiation. The color of that radiation is determined by the temperature. For example, the surface of the Sun appears yellow to us because the Sun's surface has a temperature of about 6,000 K. If the Sun were hotter it would appear bluer, if it were cooler it would appear redder. As the emitting object gets hotter, the radiation it emits shifts to shorter wavelengths. The radiation also has a well defined spectrum (light intensity of its various colors) that has a shape known as the black-body curve. One can show that, in an expanding universe, black body radiation will retain the shape of its spectrum, and appear to us as radiation of a lower temperature. The temperature of the black body radiation falls as the universe expands because the wavelength of the light shifts to longer wavelengths.

Although the cosmic background radiation was detected in 1967, its existence had, in fact, been predicted by George Gamow and collaborators in the 1940s. Gamow assumed that, at some point in its history, the universe was hot enough and dense enough for nuclear fusion reactions to take place, and predicted that the afterglow from this period should be detectable today as cosmic background radiation.

The background radiation left over from the Big Bang lets us see the universe in its early stage of evolution and gives us a window into the origin of structures that will later become the galaxies we see around us. The radiation shows us what the universe must have looked like long ago before galaxies and stars existed, it is in some sense a distant mirror. We will discuss the clues about our past revealed by the radiation in Chap. 7.

In his classic of popular science writing entitled *The First Three Minutes*, Steven Weinberg noted that the existence of this radiation could have been confirmed experimentally at the time the prediction was made. The Big Bang theory at the time was sufficiently removed from mainstream science that people did not think it worth their while to carry out experiments to confirm or deny it. The absorption of light by cyanogen molecules implied a temperature of 2.3 K for the coldest clouds of molecular gas in our galaxy. The Kelvin temperatures are quoted relative to absolute zero ($°K = °C + 273$). This 2.3 K temperature measurement dated from the 1940s and was quoted in a famous textbook in the 1950s, which stated that the temperature arrived at had a "very restricted meaning." In fact, it was an unwitting measurement of the temperature of the cosmic radiation left over from the Big Bang.

The Third Pillar of the Big Bang: The Abundances of Deuterium and Helium

> If in some cataclysm, all of scientific knowledge were to be destroyed, and only one sentence passed on to the next generations of creatures, what statement would contain the most information in the fewest words? Everything is made of atoms.
>
> *Richard Feynman*

We know that close to 90 % of all atoms in the universe are hydrogen atoms. The other 10 % are almost all helium atoms. The Big Bang theory explains why this is the case. In fact the Big Bang theory provides an accurate explanation for the relative amounts or abundances of the lighter atoms (hydrogen, deuterium, helium and lithium) that we see in the universe. How are the helium and deuterium observed? Helium was first discovered in the Sun in 1868 and later observed on Earth. Helium abundances can be determined in the spectra of hot stars, in the upper solar atmosphere, and in the solar wind. One can also indirectly infer the helium abundance by comparing theoretical model predictions for the temperatures and luminosities of stars. Deuterium abundances are even more difficult to measure. Interstellar molecules composed of hydrogen, or deuterium, carbon and nitrogen are used to estimate deuterium abundances.

As Feynman points out, "The most remarkable discovery in all astronomy is that the stars are made of atoms of the same kind as those on Earth". In 1835 the French philosopher Auguste Comte stated that "we shall never be able by any means to study the chemical composition of the stars". However, in the early 1920s, methods became available for calculating the abundances of elements in a gas by observing its spectrum. In 1925, using these methods, Cecilia Payne analyzed the spectrum of the Sun and reached the conclusion that hydrogen and helium make up 98 % of the mass of the Sun. This result surprised people who expected the composition of the Sun to be similar to that of the Earth which is made mostly of iron. The stars were thought to have the same mix of elements as the Sun, so this discovery suggested that the stars were also made mostly of hydrogen and helium. The precise conclusion is that for every 10,000 atoms of hydrogen in the universe there are 975 atoms of helium, 6 atoms of oxygen and 1 of carbon. All the other elements are present in smaller quantities than 1 atom per 10,000 hydrogen atoms. A more detailed look suggested that the abundances of the elements reflected the properties of atomic nuclei. Maybe nuclear physics processes were responsible for producing the abundances of the elements.

Already in 1903 Rutherford and Soddy had hinted that nuclear processes were responsible for generating energy in the Sun;

> The maintenance of solar energy, for example, no longer presents any fundamental difficulty if the internal energy of the component elements is considered to be available, that is, if processes of subatomic change are going on.

When this question was examined by von Weizacker he found that many different sets of conditions were required to produce the mix of elements that we see in nature.

The Sun produces energy by turning hydrogen into helium. So, why can't nuclear fusion in stars be used to account for all the helium in the universe? There is not enough time. The stars can only account for about 2 % of the helium production in the universe. To create elements from hydrogen requires extremely high temperatures. If the stars cannot do the job, where can we find another furnace hot enough to form helium?

George Gamow was the first to explore the idea that nuclear processes could have taken place in the first few minutes of the Big Bang when it was hot enough for nuclear fusion to occur. The physicists Alpher and Herman worked out the details and found that nuclear processes which we will discuss below produced just the right abundance of helium to match that seen in stars. These processes also account for lithium and deuterium abundances. The inescapable conclusion is that only the very lightest atoms were created during the hot Big Bang. The heavier elements it turns out are made inside stars and during supernova explosions.

To understand more of the workings of nuclear alchemy we need to explore the properties of atoms. Atoms are made from particles called electrons, neutrons, and protons. The neutrons and protons form the nucleus, and the electrons orbit the nucleus. In this sense, the atom resembles the solar system. In the solar system, most of the mass resides in the Sun; in atoms most of the mass resides in the nucleus. There are certain allowed orbits or energy levels for electrons in atoms. These can be precisely calculated using the formalism of quantum theory, which was developed in the 1920s. The electron can jump from one level to another by emitting or absorbing light of known energy and wavelength (i.e. color). This is a crucial property of atoms, which astronomers make use of to measure redshifts as we have seen. Without sending a probe to the Sun, we can tell which elements are present in the Sun. The absorption lines in the Sun's spectrum occur because atoms near the Sun's surface absorb some of the light generated in the Sun's interior. The absorption occurs because electrons in these atoms jump up from one energy level to another and absorb light in the process.

You are no doubt familiar with the concept of chemical elements, such as carbon, nitrogen, and so on. What differentiates one element from another is the number of protons in the atomic nucleus. Hydrogen, the simplest atom, consists of one electron orbiting one proton. The helium atom consists of two protons and two neutrons in the nucleus with two electrons in orbit. The number of protons in the nucleus determines how many electrons are present in each atom. This is because atoms have no net charge; the negative charge of the electrons is needed to balance the positive charge of the protons. It is the properties of the electron orbits that determine the chemical properties of the elements. The periodic table of the elements is an ordering of elements according to the number of protons in their nucleus which in turn determines their chemical properties.

It is a triumph of physics to have explained the properties of the periodic table in terms of the structure of atoms. This structure is determined by some simple fundamental equations. The power of physics lies in the ability to unify seemingly unrelated phenomena using simple underlying principles. A number of key discoveries about atoms and their constituent particles were made at the Cavendish Laboratory in Cambridge, England. The discovery of the electron, the neutron, and the atomic nucleus were all made at the Cavendish Laboratory. Ernest Rutherford, who discovered the atomic nucleus, was once asked how it was that he always seemed to be riding the crest of the wave in nuclear research. His answer was characteristic "I created the wave." Immodesty aside, this is an attribute of great men, they create the environment in which their and other talents can flourish. Niels Bohr did this to great effect in Copenhagen in the 1920s and 1930s. Many great physicists of that period passed through Bohr's institute.

The Particle Zoo

To understand more of the early universe we need to briefly discuss the elementary particles. These particles are classified according to which forces of nature they respond to. We have so far encountered two properties of particles; charge and mass. Another property is spin. Spin can be measured and has whole number values (for example 1, 2) or fractional values (such as $\frac{1}{2}$ and $\frac{3}{2}$). Particles such as electrons have a spin of one half and are called fermions. Particles with integer spins are called bosons. Bosons and fermions behave in different ways when they are pushed close together.

Since like charges repel, and the atomic nucleus contains only positively charged particles and neutral particles, what holds the nucleus together? A force of attraction called the strong force overcomes the electric repulsion on small scales. We have a picture of the atom with an outer shell of electrons and a compact nucleus. The atomic nuclei can also react with each other. They can stick together in a process known as nuclear fusion. For elements lighter than iron, the fusion process releases energy, although it takes very high temperatures to get it going. It is this energy release that powers the Sun and most stars. The reason fusion reactions take place only at very high temperatures is that nuclei are positively charged and therefore repel each other. They have to approach each other at high speeds to overcome the repulsion and get close enough for the short range strong interaction that causes fusion to be felt. One might think of rolling a bowling ball up the side of a hill with a crater at the top. If the ball has enough speed if can make it to the top and settle in the crater. If the speed is too slow the ball will roll back down the hill. Temperature is a measure of the speed of particles hence the need for high temperatures for fusion to take place.

Another force, the weak force, plays a role in fission. We have now encountered the four fundamental forces of nature, the strong force, the weak force, electromagnetism, and gravity. Electromagnetism describes the phenomena arising from electric and magnetic forces which were united in one framework by

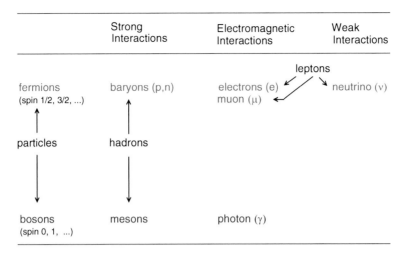

	Strong Interactions	Electromagnetic Interactions	Weak Interactions
		leptons	
fermions (spin 1/2, 3/2, ...)	baryons (p,n)	electrons (e) muon (μ)	neutrino (ν)
particles	hadrons		
bosons (spin 0, 1, ...)	mesons	photon (γ)	

FIG. 2.4 The particle terminology. For each particle its strongest interaction is shown. In general it has all the interactions to the right of its entry, so that baryons, for example, have electromagnetic and weak interactions in addition to their strong interaction. Fermions are shown in *green* and bosons are shown in *blue*

James Clerk Maxwell. He showed that electromagnetic fields could propagate through space in the form of radiation that we see as light.

The fermion family of particles can be further subdivided into hadrons and leptons. The hadrons are strongly interacting fermions, whereas the leptons are weakly interacting fermions. Examples of leptons are electrons and neutrinos; examples of hadrons are protons and neutrons (protons and neutrons have spins of one half). It turns out that hadrons have structure. We believe them to be made of even more fundamental particles called quarks. Hadrons which consist of three quarks, are called baryons. Hadrons which consist of a quark antiquark pair, are called mesons. Antiquarks are an example of antimatter. Each particles has its own antimatter counterpart, a particle with opposite charge but identical mass. The electron's counterpart is called a positron. The existence of the positron was predicted by Paul Dirac (1902–1984) who devised a theory combining quantum physics and relativity.

The particle terminology we have introduced is shown schematically in Fig. 2.4. Baryons will be the subject of further discussion and are the most common form of hadron. To keep things simple, we may think of neutrons and protons when we use the word baryon. Other baryons exist, but they are very-short lived particles and do not play a role in our story.

Is there such a thing as an elementary particle? The atoms were shown to have structure, then the nucleus was shown to have structure, and most recently, the constituents of the nucleus have been shown to have structure. We can ask ourselves whether a fundamental level of matter exists. Do the quarks themselves have structure? We cannot answer this question at the present time. We also cannot explain the masses of the particles. Experiments underway at the Large Hadron Collider in Geneva, Switzerland aim to test theories of the origin of particle masses.

The First Three Minutes

Calculations of the fusion rates in the early universe predict that after nucleosynthesis is complete, the atomic matter in the universe should consist of 75 % hydrogen by mass, with 25 % helium, and only traces of other elements. This remarkable prediction is in agreement with the observed chemical composition of the atmospheres of the oldest known stars. Let us take a closer look at the process of helium formation.

Initially the universe was so hot that neither neutral atoms nor even atomic nuclei could exist. At that point the universe consisted of a mix of radiation and particles as shown in Fig. 2.5. When the universe was about 2 min old, the temperature had 'cooled' to one billion Kelvin, much hotter than the center of the Sun. The density of the universe at this time is about half that of the air you breathe.

When the universe had cooled down to one billion degrees, it was cool enough for deuterium to hold together. This is because at temperatures higher than this a proton colliding with a deuterium nucleus has enough energy to break it apart. This illustrates the general point that at lower temperatures, matter exists in bound structures, while at higher temperatures, matter exists in the form of individual particles flying around (see Fig. 2.6).

The presence of deuterium enables helium to form at a rapid pace. The deuterium nucleus consists of one proton and one neutron. It is an isotope, having the same chemical properties as

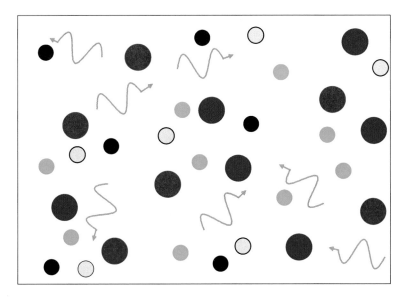

FIG. 2.5 Less than 1 s after the Big Bang, densities are high and interactions happen quickly. Protons (*red*) can convert to neutrons (*blue*) in reactions involving positrons (*yellow*) electrons (*brown*) and neutrinos (*green*). Light particles (photons) are shown as *orange* wiggly lines. The reactions take place via the weak interaction, one of the fundamental forces of nature

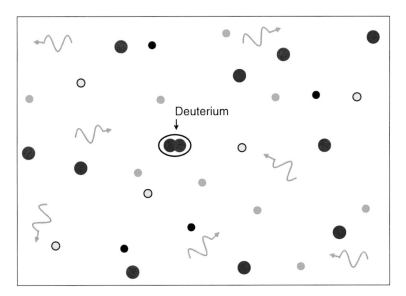

FIG. 2.6 As the universe cools interactions freeze out. Residual neutrons (*blue*) combine with protons (*red*) to form Deuterium, Helium, and Lithium in the first few minutes after the Big Bang

hydrogen, but with a nucleus that contains an extra neutron. We can write the formation of deuterium as an equation:

$$n + p \leftrightarrow D + \gamma.$$

That is to say, a neutron (n) plus a proton (p) combine to form deuterium (D) and release energy in the form of light (γ). The last symbol on the line, γ, represents a photon or light particle. The reaction can go either way resulting in the formation or destruction of deuterium.

Once deuterium has formed, two particles of deuterium can combine to form a helium nucleus. There is a small window of opportunity for helium to form in the early universe. It must be cool enough that deuterium nuclei can survive, but hot enough that the deuterium nuclei can collide and form helium. It is remarkable that in the 14 billion years of the history of the universe there were a few minutes during which conditions were just right for nuclear fusion to take place. Fusion reactions don't take place at room temperature because of the force of repulsion between positively charged atomic nuclei. As we described in our hill with a crater metaphor, the repulsive force must be overcome for the particles to get close enough together to feel the strong nuclear force, which is a short range force. It is only at high temperatures that nuclei collide with sufficient speed to get close enough to interact via the strong force.

Deuterium and the Formation of Helium

The formation of deuterium is critical to the formation of helium because a helium nucleus can form by the collision of just two deuterium particles. To form a helium nucleus spontaneously from protons and neutrons would involve four particles colliding at the same time. Accidents involving two cars occur much more frequently than accidents involving four cars. So it is with particle collisions.

Almost all the deuterium formed in the Big Bang is used to make helium nuclei. To calculate how much helium formed in the big bang we need to know how much deuterium was present in the early universe. This, in turn is determined by the number of neutrons relative to protons at the time of nucleosynthesis.

Neutrons are neutral particles that have a mass slightly larger than that of the proton. A free neutron can decay into a proton and an electron through the following reaction:

$$n \rightarrow p + e + \bar{\nu}$$

The symbol $\bar{\nu}$ stands for an antineutrino. The typical time for this reaction to take place is about 15 min. This is the time it takes for a free neutron in space to turn into a proton. Note that a neutron inside a nucleus is stable, it will not turn into a proton. It may seem strange to see this number that we use in everyday life being relevant to the early universe. Initially there are roughly equal numbers of protons and neutrons in the universe. The neutrons can turn into protons but there is a resupply of neutrons from protons colliding with neutrinos. As the temperature drops there are fewer neutrons relative to protons. By the time the universe is a few seconds old, proton-neutron interchanging reactions have been brought to an end because the density is low enough that neutrinos cease to interact with anything. After 2 min the universe is cool enough for the remaining neutrons to combine with protons to form deuterium. At this time there are 2 neutrons for every 14 protons in the universe. These neutrons are absorbed into deuterium, so there are 2 deuterium nuclei for every 12 protons in the universe, Almost all these deuterium nuclei then combine to form helium nuclei. The end result is that there is 1 helium nucleus for every 12 protons in the universe which amounts to one quarter of the mass of nucleons in the universe.

A helium nucleus consists of two protons and two neutrons. Essentially, all the neutrons in the universe available at the time of nucleosynthesis end up inside helium nuclei. The amount of helium in the universe is thus determined by the neutron-to-proton ratio at the time when deuterium forms. For example if the neutron-to-proton ratio had been zero,– that is, no neutrons– no helium would form at all. If the neutron to proton ratio had been one, then the universe would consist entirely of helium. The helium abundance is a strong prediction of the Big Bang theory. There is not much room for maneuver. If the universe started in a hot, dense phase, then one quarter of the mass of baryons must consist of helium nuclei. The fact that helium was produced in the Big Bang would also explain why the helium abundance does not vary much from one location to another in the universe.

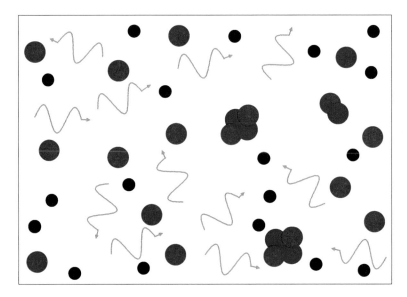

FIG. 2.7 In the first 400,000 years after the Big Bang, temperatures are so high that electrons (*brown circles*) cannot join protons (*red circles*) or helium and deuterium nuclei to form atoms. This is because the photons (*orange wavy lines*) are energetic enough to knock eletrons away from atoms. During this period photons cannot travel far before interacting

After nucleosynthesis, the universe consisted of a mix of photons, electrons, protons, and helium and deuterium nuclei (Fig. 2.7). It was still much too hot at this time for atoms to form.

There is just a little deuterium left over from the Big Bang because the production of helium does not completely use up all of the deuterium nuclei. We can use the abundance of deuterium seen today to estimate the density of baryons in the universe. For a higher baryon density there should be very little deuterium left, but for a lower baryon density, more deuterium should survive. The abundance of deuterium in the universe today implies a present baryon density of the universe of 0.2 atoms per cubic meter, about double the density of matter seen in stars and gas. Studies of the cosmic background radiation lead us to the conclusion that the total density of matter in the universe is 1.3 atoms per cubic meter. Putting together the two we are forced to conclude that most of the matter in the universe does not consist of ordinary atoms. The question of the nature of this non-baryonic dark matter in the universe is one of the major issues of cosmology.

The Nature of the Big Bang

Most cosmologists explain their observations within the framework of the Big Bang theory. Some theoreticians like to think about what happened at the earliest times, minute fractions of a second after the Big Bang, and even discuss how the Big Bang might have occurred. The theory of inflation purports to account for the starting conditions of the Big Bang. In particular it explains why the universe is expanding, why it is homogeneous, and, why on large scales the universe has a flat geometry. Variants of the theory also suggest that different parts of the universe each have their own elementary particles and constants of nature. These areas form a sort of foam with a bubble structure. This idea, known as the multiverse, purports to explain why the constants of nature seem fined tuned so that we can exist in the universe. For example if the strong interaction was slightly stronger, then two protons would be able to bind together and form helium without two neutrons. This would have happened in the early universe with the result that no hydrogen would remain to provide fuel for stars, and water could not exist. These coincidences have led the distinguished physicist Freeman Dyson to state that:

> As we look out into the universe and identify the many accidents of physics and astronomy that have worked together to our benefit, it almost seems as if the universe must have known that we were coming.

In the multiverse picture, the constants of nature take a wide range of values. Intelligent observers exist only in those rare bubbles in which by pure chance the constants happen to be just right for life to evolve. These ideas are very speculative and make few predictions that we can test. Recently though, it has been suggested that these processes might leave their mark on the cosmic background radiation in a way that we could detect. An interesting prediction of the multiverse idea was that galaxies would form in regions where the dark energy density is comparable to the dark matter density, this is indeed the case in our observable universe and could be viewed as a prediction of the model since it predates the discovery of dark energy.

It has also been argued that the universe could have spontaneously been created out of nothing. In quantum theory any process which is not forbidden by the laws of nature has a finite probability of taking place. In this view there is no cause needed for the creation of the universe.

The Timeline of the Universe

The Big Bang theory states that the expansion of the universe began at a finite time in the past, in a state of enormous density and pressure. As the universe grew older it cooled and various physical processes came into play which produced the complex world of stars and galaxies we see around us. The history of the universe can be outlined as follows:

The earliest period that has any significance in cosmology is known as the Planck time. This amazingly small time interval is 10^{-43} of a second. After this time, general relativity can be used to describe the interaction of matter and radiation with space. Before the Planck time, we have no theory to describe the universe. We require an idea that incorporates the concepts of quantum physics and general relativity into one unified theory. Stephen Hawking and his colleagues work on such problems, but no definitive answer has been arrived at yet.

- 0 to 10^{-43} s: This takes us from the moment of the big bang up to the Planck time. We have no physical theory to describe the behavior of matter under the conditions that prevailed this early in the history of the universe.
- 10^{-43} to 10^{-35} s: During this time a slight excess of matter over anti-matter was produced. After the matter and antimatter annihilated, a small amount of matter was left over. Today there is only one baryon per billion photons in the universe.
- 10^{-35} to 10^{-6} s: The fundamental forces separate into four recognizable forces that we see today. At the end of this era, quarks combine to form hadrons (e.g. neutrons and protons) and mesons.
- 10^{-6} to 10^{-4} s: This is known as the hadron era. During this time, the baryons and antibaryons annihilated, resulting in a slight excess of baryons being left over to form the stars and galaxies that we see around us.

FIG. 2.8 Four hundred thousand years after the Big Bang, the universe has cooled sufficiently that atoms can form for the first time and light can travel freely through the universe. We detect this light today as the cosmic background radiation. The atoms present at this time are almost all hydrogen and helium

- 10^{-4} to 10 s: This is the lepton era. Leptons are particles that feel the weak interaction, such as the electron. At 0.1 s the neutron to proton ratio starts to tilt in favor of protons. At 1 s, neutrinos stop interacting with matter and each other. At 10 s the neutron to proton ratio became fixed, which in turn determined the deuterium and hence the helium abundance in the universe.
- Three to twenty minutes: Nuclear fusion occurs, producing helium and a little deuterium and lithium. The explanation of the helium abundance is one of the triumphs of the Big Bang theory. The big bang also explains the abundance of very small amounts of deuterium and lithium.
- 10 to 10^{11} seconds (3,000 years): This is the radiation era. Radiation dominates the energy density of the universe during this period.
- 10^{13} s (400,000 years): The universe becomes cool enough for neutral atoms to survive and thus become transparent. Radiation can travel freely though the universe with very little chance of being scattered or absorbed (see Fig. 2.8).

- 10^{13} s to present: During this time, regions of the universe that are slightly denser than their surroundings begin to collapse and eventually form stars, galaxies and clusters of galaxies. With the formation of the first stars nuclear fusion reactions occur again inside these stars for the first time since the first few minutes of the universe.

The Evolution of the Universe

As the universe ages it evolves. From being hot enough to act as a furnace for nuclear reactions, the universe has cooled to a few degrees above absolute zero. From the inferno of the first 3 min the universe cooled to a temperature that allowed the first clouds of primordial gas to collapse, at which point the first stars formed. The focus of the largest ground based 8–10 m mirror telescopes and space telescopes is to get a direct window into the first billion years of the universe and observe the growth and evolution of the cosmic structures mapped out by stars and galaxies. Astronomers are currently discovering objects that are recognizable as galaxies at a redshift of about eight. The light from these objects was emitted when the universe was less than 5 % of its present age (see Fig. 2.3).

By looking over great distances we look back to ever earlier times. To understand our findings we make the assumption that, averaged over sufficiently large scales, the universe is essentially the same everywhere. We thus assume that the galaxies we see in the distant past resemble what galaxies in our local 'neck of the woods' would have looked like at comparably early times. The practical challenges are; the most distant galaxies are extremely faint, their light gets shifted to progressively longer wavelengths and, dust and gas can obscure the light from young stars.

We shall describe this fascinating world of galaxies, the realm of the nebulae as Hubble called it, in Chap. 3.

Further Reading

The First Three Minutes. S. Weinberg. New York: Basic Books, 1993.

Just Six Numbers. M. Rees. New York: Basic Books, 2000.

The Magic Furnace: The Search for the Origins of Atoms. M. Chown. Oxford: Oxford university Press, 2001.

The Origin of the Chemical Elements. R. J. Tayler and A. S. Everett. London: Wykeham, 1975.

Seeing Cosmology Grow. P. J. E. Peebles. Annual Reviews of Astronomy and Astrophysics, 2012

3. The Visible Universe

All humans are brothers. We came from the same supernova.

Allan Sandage

For an astronomer, there are really two forms of matter; visible and invisible. Visible matter emits electromagnetic radiation, which travels through space and is detected with telescopes. Invisible matter does not produce detectable amounts of radiation. The presence of invisible, or dark matter, can only be inferred through its gravitational effect. The study of visible matter enables us to infer the distribution of the dark matter whose gravity was responsible for the formation of the first stars and galaxies. In this chapter we explore the visible matter as seen in stars, galaxies and gas.

The Lives of the Stars: Birth

Galaxies such as the Milky Way are continuously forming stars. Stars form out of the very tenuous (by earth standards) gas clouds that exist between the stars. Known as molecular clouds, these have temperatures of 10–30 K, sufficiently low that hydrogen can exist in molecular form. Interestingly, molecular clouds contain other molecules, such as carbon monoxide, water, ethanol, ammonia, and even amino acids. The existence of organic molecules in the interstellar medium suggests the possibility that life may have originated in molecular clouds in interstellar space.

The inward acting force of gravity is countered by the gas pressure in these clouds. Star formation can be triggered in molecular clouds by a sudden increase in density. This may be caused by the formation of stars nearby; star formation can spread like wildfire once it gets going. Figure 3.1 shows a nearby star-forming region with young stars and gas clouds.

G. Rhee, *Cosmic Dawn: The Search for the First Stars and Galaxies*, Astronomers' Universe, DOI 10.1007/978-1-4614-7813-3_3, © Springer Science+Business Media, LLC 2013

FIG. 3.1 Undulating bright ridges and dusty clouds cross this close-up of the nearby star-forming region known as the Lagoon Nebula. A color composite of visible and near-infrared data from the 8-m Gemini South Telescope, the entire view spans about 20 light-years. The cosmic Lagoon is found some 5,000 light-years away toward constellation Sagittarius and the center of our Milky Way Galaxy (Credit: Julia I. Arias and Rodolfo H. Barbá, (Dept. Fisica, Univ. de La Serena), ICATE-CONICET, Gemini Observatory/AURA)

Molecular clouds rotate, such that when clouds start to collapse, a disk forms. It is out of such a disk that our own solar system is believed to have formed. These disks have been observed, associated with jets of matter flowing along the rotation axis of protostars.

In the initial stages of their lives, stars shine very bright. Since they are radiating away a lot of their energy, they can shrink in size, releasing more energy and heating up the core of the star. Eventually, the temperature rises to the point that nuclear fusion reactions begin in the star's core. The central temperature of the star continues to rise, as do fusion rates, until a balance is achieved. That is to say, the energy generated in the star's core becomes equal to the energy sent into space at the surface of the star.

Brown Dwarfs and Planets Outside the Solar System

The minimum mass required for a contracting gas cloud to turn into a star is about one tenth of the mass of the Sun. These stars are called red dwarfs, they burn their hydrogen so slowly that they can keep nuclear fusion going for 100 times longer than the Sun, about 1 trillion years. Clouds of gas having a mass less than one tenth of the Sun's mass do not heat up sufficiently in their core for nuclear reactions to start. Such objects are known as brown dwarfs. Some examples of brown dwarfs have been found with the Hubble Space Telescope. Because brown dwarfs are so faint, it is difficult to estimate how many of them exist in our galaxy. We do know that less massive stars occur much more frequently than more massive stars. If we extrapolate this trend to brown dwarfs, the numbers of these objects could be quite high.

It is not clear what fraction of stars have their own planetary systems. Discoveries of several hundred nearby planetary systems suggest that the formation of planets surrounding stars may be a fairly frequent phenomenon.

The History of Star Formation in the Universe

How can we detect the birth of stars in other galaxies? Newly born stars are still surrounded by clouds of gas, and dust, which absorbs the light from the young stars and re-emits it at infrared wavelengths. Our galaxy re-emits about a third of the light from stars as infrared radiation. Telescopes such as the European Space Agency's Herschel telescope (Fig. 3.2) are providing a new view of star-forming regions in our and other galaxies.

We can estimate the amount of star formation taking place by measuring the infrared light from galaxies. Star-forming regions have hot young stars associated with them. These stars emit ultraviolet radiation. We can use the amount of ultraviolet light emitted per unit volume to estimate the star formation rate per unit volume. We must also take into account the fact that ultraviolet light is scattered by dust in galaxies.

We can measure star-forming activity in nearby galaxies, but we can do the same with distant galaxies and compare the results. Are distant galaxies forming stars at a faster rate than nearby galaxies or has the rate of star formation been fairly constant over time? The answer is shown in Fig. 3.3 which depicts the star formation history of the universe. The rate of star formation increases steeply during the first billion years, peaks about 1 billion years after that, and then gradually decreases until the present day. It is remarkable that we can measure how the number of stars per

FIG. 3.2 Front view of the Herschel Satellite. The first observatory to cover the entire range from far-infrared to sub-millimeter wavelengths and bridge the two. The satellite was launched 14 May 2009 on an Ariane rocket from Kourou French Guyana (Credit: ESA/AOES Medialab)

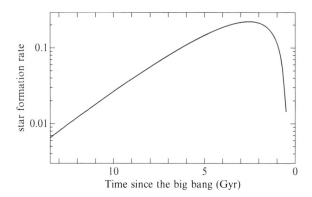

FIG. 3.3 The cosmic star formation history derived from the ultraviolet luminosities of galaxies. The rate at which a typical region of space formed stars increased rapidly during the first 2 billion years of the universe and then started a slow decline to the present day. It is during the first 2 billion years following the big bang that galaxies such as our Milky Way came into existence. The units of the horizontal axis are Gyr, short for Gigayear which is a billion years

year being born in a volume of the universe has changed since the big bang. We have come a long way in the 400 years since the telescope began to be used as an astronomical tool!

The Lives of the Stars: Middle Age

Stars are massive spheres of hot gas that glow because of their high temperatures. The nearest middle-aged star to us is the Sun, a rather average star in our galaxy. The radius of the Sun is about twice the Earth-Moon distance. The mass of the Sun is about 300,000 times the Earth's mass. If we could capture 1 s worth of the total energy emitted by the Sun, we could fill the energy needs of the human race for the next million years. The average density of the Sun is about equal to that of water.

The composition of the Sun is 28 % helium, 70 % hydrogen and 2 % heavier elements. The surface temperature of the Sun is about 6,000 K or about double the temperature of a blowtorch. The temperature near the center of the Sun is about 15 million degrees Kelvin. It was only in the 1930s that the true source of the Sun's energy was discovered. We pointed out in Chap. 2, that the center of the Sun is hot enough that nuclear fusion reactions can take place. Keep in mind though that 90 % of the helium that is present in the Sun was produced in the early universe.

The fusion of hydrogen into helium releases energy, and it is this energy that powers the Sun and makes it shine. The fusion process takes place in three steps. The net effect of these three steps is that four protons fuse together to produce a helium nucleus and some radiation. This fusion process does not take place at room temperature because of the repulsion force between protons. The strong force, which holds atomic nuclei together, has a very short range. Two protons must collide with very high energy in order to get close enough to each other to feel the strong force. This only happens at very high temperatures.

Since a helium nucleus is less massive than four protons, the mass difference must be accounted for. The mass difference is, in fact, converted to energy, following Einstein's famous $E = mc^2$ equation. The mass difference is a very small fraction, 0.7 % of the original mass. When 1 kg of hydrogen fuses into helium, 993 g of helium are produced, while 7 g of mass turn into energy. Four million tons of mass per second are converted to energy in the Sun. The Sun is, in effect, a controlled hydrogen bomb. It is amazing that the laws of nature have given rise to nuclear fusion reactors capable of maintaining a stable energy output over billions of years.

We know how much energy the Sun produces per second, and we know the mass of the Sun, so we can calculate how much longer we can expect the Sun to keep on shining. If the Sun remains as it is until it has used up all of its hydrogen supply, it will have enough fuel to keep shining for another 70 billion years. In fact the Sun cannot use up all its hydrogen because it is only in the core that it is hot enough for fusion to happen. Calculations show that the Sun will change in structure after using up about 13 % of its total store of hydrogen. This should take about 9 billion years. Since the Sun has already been around for 4.6 billion years, it is already in middle age.

The fact that stars like the Sun shine by converting hydrogen into helium together with the fact that stars have a finite mass and thus a finite supply of fuel suggests that stars evolve. Stars are born and go through several stages before they reach the end of their energy producing "lives". The Sun is not an extreme star. Its properties such as mass and size are near the middle of the properties that are measured for stars in our galaxy.

The Lives of the Stars: Mid-Life Crisis!

In a few billion years the hydrogen in the Sun's core will be converted to helium. The core of the Sun will then cease to generate energy, and start to shrink and heat up. Surprisingly, perhaps, while the core of the Sun undergoes this, the outer layers do just the opposite. They expand and cool down. The result is that, during this phase, a star like the Sun appears to get redder and larger, becoming what is known as a red giant. During the red giant stage, the Sun will engulf the inner planets and vaporize any life remaining on earth. If any humans are still around, they had better plan on finding a home on another planet or moon farther out from the Sun. In the red giant stage, the Sun will start fusing hydrogen into helium in a shell surrounding its core. Meanwhile, the core shrinks and gets hotter. When the core temperature reaches 100 million degrees, it becomes hot enough for helium fusion to take

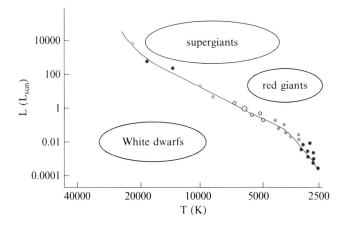

FIG. 3.4 The Hertzrpung-Russell diagram. Each point represents a star. The star's surface temperature is plotted against its luminosity. An L value of 1 in these units is the luminosity of the Sun. Our Sun is shown as the larger *yellow circle*. The points are color coded to make the point that stars with low surface temperatures appear *red* whereas hot stars appear *blue* in the night sky

place, producing a heavier element, carbon. We are born of stars in that all the carbon in our bodies was manufactured inside them. During the red giant stage, stars can lose substantial amounts of mass and become surrounded by clouds of their own gas. The Hubble Space Telescope has taken some stunning images of these clouds.

Figure 3.4 is a plot of surface temperature against energy output (or luminosity as we call it) for several stars. We can see that the range of star luminosities found in nature is much larger than the range of surface temperatures. When the stars are in the hydrogen burning phase of their lives, they occupy a diagonal line in this diagram. We refer to this line as the main sequence. The main sequence is ordered by mass such that the most massive stars are hot, blue and of high luminosity while the least massive stars are cool, red and have low luminosity. When a star becomes a red giant it leaves the main sequence and becomes more luminous and redder in color. Massive stars are known to have shorter lifetimes, so for a cluster of stars all born with different masses, but at the same time, the main sequence will appear to peel off. The point at which stars are leaving the main sequence is known as the turn-off point. The turn-off point is a measure of the age of the star cluster. Some of these clusters are very old, between 11 and 13 billion years.

These ages give us an estimate of the age of the galaxy since it must be older than the oldest objects in it. The oldest stars in our galaxy have very little iron in them since they were formed from gas that had not been polluted so to speak by supernova explosions. We have found a 13 billion year old star in our galaxy that has 100,000 times less iron than is seen in the Sun. Since there actually is some iron in this star it cannot be one of the first stars to form after the big bang. We are still searching for the first stars.

Certain stars during the red giant phase can become unstable and start pulsating. These are known as variable stars since the pulsation results in periodic variations in brightness. The pulsation time can vary from hours to years, depending on the kind of star and its mass. We can use these variable stars to estimate their distances from us. This is why Hubble was so excited to discover a variable star in the Andromeda nebula (see Chap. 1). It meant he could calculate the distance to that nebula and establish that it was a galaxy like our milky way. We are fortunate that these stars are so bright. A star like our Sun would not be detectable at the distance of the Andromeda galaxy.

Rest in Peace: The Death of Stars

The life of a star is a battle between the forces that generate heat in the stellar interior and the force of gravity, which wants to crush matter ever inward. After stars run out of fuel, they reach the final stages of their lives. The masses of stars determine their ultimate fate. Stars less massive than 1.4 solar masses end their lives as white dwarfs. After the red giant phase, the Sun will shrink to the size of the Earth and stop generating energy. It will shine with a bluish glow but will be no brighter in the Earth's sky than the full moon today. The only reason white dwarfs shine is that they have a high temperature, but eventually they will cool because the energy they radiate is not being replaced. The gravity on the surface of a white dwarf is very strong. If you dived off a 30 ft platform into a swimming pool on a white dwarf, it would take you about 1/100 of a second to hit the water compared to about 1.4 s on earth. The matter in a white dwarf has a crystalline structure, but what stops the white dwarf from collapsing?

An effect called the exclusion principle provides an effective counter to the crushing force of gravity. Electrons do not "like" to occupy the same position when they have the same velocities. A consequence of this is that it costs energy to shrink a star like the Sun beyond the size of the Earth, and a balance point is reached. A sugar cube's worth of a white dwarf weighs about a ton. A lump the size of beach ball weighs as much as an ocean liner.

Combining quantum physics and the theory of relativity an Indian physicist Subrahmanyan Chandrasekhar predicted in 1930, that a white dwarf having a mass larger than 1.4 solar masses, will collapse to form a neutron star. The calculation was carried out on a ship sailing from India to England where the 20 year old Chandrasekhar was to begin graduate study in physics at Cambridge University. A neutron star has a radius of a few miles, and a teaspoonful of such a star weighs as much as a battleship. Neutron stars have their own upper mass limit. Stars more massive than five solar masses or so will end their lives as black holes–objects from which nothing, not even light itself, can escape. White dwarfs, neutron stars, and black holes are the three possible end states of stellar evolution.

The key to understanding stellar collapse lies in the fact that the most stable atomic nucleus is iron. Nuclear energy can be released from lighter elements than iron by fusing them together. However, no nuclear energy can be released by splitting or adding to an iron nucleus. As stellar evolution progresses, massive stars find themselves with increasingly heavier elements in their centers. Finally when a star contains an iron core, no more energy release through fusion is possible, collapse is bound to follow. The collapse happens in less than a second and produces a neutron star. During the collapse, the neutron star bounces slightly. The bounce expels matter at tremendous speeds. During the formation of the neutron star huge numbers of neutrinos are produced, which contribute to the energy of the explosion.

In 1934, two astronomers working at Caltech, Walter Baade and Fritz Zwicky, suggested that the transition of an ordinary star into a neutron star consisting mainly of neutrons could be accompanied by a tremendous explosion known as a supernova. Zwicky is a legendary figure among astronomers. Known for his eccentricity in a field of eccentrics, he came up with a remarkable number of original ideas, including the notion of dark matter in galaxy clusters. The formation of a neutron star should produce a huge flux of neutrinos. It was a triumph of astrophysics when the neutrinos from a supernova were first observed. In 1987, a supernova was seen to go off in the Large Magellanic cloud, a companion galaxy to the Milky Way. Luckily for astrophysicists, two neutrino detectors, one in Ohio and one in Japan, were operational at the time. About 18 h before the supernova was first observed, the detectors caught 19 neutrinos during a 12 s interval. This was an amazing confirmation of the concept of neutron star formation proposed by Zwicky and Baade 53 years earlier. The existence of pulsars, spinning neutron stars, was discovered back in 1967, but no neutrinos had ever been detected from a supernova explosion. There is a PBS NOVA special on the discovery of supernova 1987a as it is called which really conveys through interviews the sense of wonder and enthusiasm that scientists felt at that discovery.

Fusion in the cores of stars converts hydrogen into helium, helium into carbon, and, for the most massive stars silicon and iron, but there the sequence ends. The elements such as copper, gold and tin that we see around us were created in supernovas. As James Dunlop puts it

> The past history of star-formation activity even affects today's financial markets, with the seeming ever rising price of rare commodities such as gold being due, in large part, to the rarity and brevity of the violent supernova explosions in which all gold was originally forged.

Supernovas: The Ultimate Cosmic Fireworks

During the supernova phase, a star can become one billion times more luminous than the Sun, and for a short time outshine an entire galaxy. Supernovas are characterized by the way they brighten and fade and by their spectra. Supernovas as classified into several categories; Type Ia, Ib Ic and Type II. Of primary interest to cosmologists are Type Ia supernovas because they all reach a similar maximum luminosity. Type Ia supernovas occur by the thermonuclear explosion of a white dwarf star. These white dwarfs accrete mass from a companion star until they exceed their mass

FIG. 3.5 The Crab Nebula, the result of a supernova seen in 1054 A.D., is filled with filaments. This image was taken by the Hubble Space Telescope. The Crab Nebula spans about 10 light-years. In the nebula's very center lies a pulsar: a neutron star as massive as the Sun but with only the size of a small town. The Crab Pulsar rotates about 30 times each second (Credit: NASA, ESA, J. Hester and A. Loll (Arizona State University))

limit and a thermonuclear explosion occurs. Other supernovas are caused by iron core collapse of a massive star.

The oldest records of supernovas come from China. In the year A.D. 1054, a supernova was observed in the constellation Taurus and recorded by Chinese astronomers. A star in that constellation became so bright that it was visible even in daylight. It could be seen with the naked eye for a few weeks following its peak in brightness. We see today at the location of this star an expanding cloud of gas known as the Crab Nebula (Fig. 3.5).

FIG. 3.6 These images show Type Ia supernova PTF 11kly, the youngest ever detected, over three nights in August 2011. The *left image* taken on August 22 shows the location before the supernova went off. The *center image* taken on August 23 shows the supernova at about 10,000 times fainter than the human eye can detect. The *right image* taken on August 24 shows that the supernova is six times brighter than the previous day (Credit: Peter Nugent and the Palomar Transient Factory)

The nebula consists of the outer layers of the star that were blasted into space at enormous speeds. Following a supernova event expanding clouds can be visible for thousands of years before fading from view. At the center of the Crab nebula lies a neutron star 15 miles in diameter that rotates over 30 times a second. This neutron star formed during the supernova event. The first supernova seen in a galaxy other than our own was observed in 1885 in the Andromeda nebula. Many supernovas have been observed in external galaxies since then. The recently discovered supernova in M101 also known as the Pinwheel Galaxy is shown in Fig. 3.6. The star was too faint to be detected in images taken with ground based telescopes and the Hubble Space Telescope prior to the explosion. The last supernova observed in our own galaxy was seen in 1604.

In a typical galaxy we expect to see a supernova every 50 years. With modern digital detectors we can survey large areas of the sky and check for supernovas automatically. In 2011 290 supernovas were detected in this way.

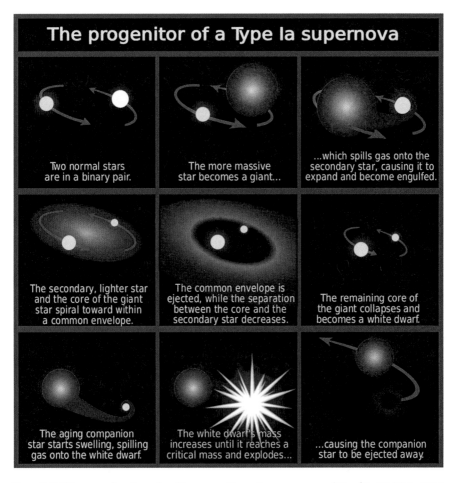

The progenitor of a Type Ia supernova

Two normal stars are in a binary pair.	The more massive star becomes a giant...	...which spills gas onto the secondary star, causing it to expand and become engulfed.
The secondary, lighter star and the core of the giant star spiral toward within a common envelope.	The common envelope is ejected, while the separation between the core and the secondary star decreases.	The remaining core of the giant collapses and becomes a white dwarf.
The aging companion star starts swelling, spilling gas onto the white dwarf.	The white dwarf's mass increases until it reaches a critical mass and explodes...	...causing the companion star to be ejected away.

FIG. 3.7 The mechanism leading to a Type Ia supernova (Credit: NASA, ESA and A. Feild (STScI))

Supernovas as Cosmological Tools

What kinds of stars end their lives as supernovas? It is believed that any star born with a mass of more than 8 solar masses will end its life as a supernova. Such supernovas have a wide range of peak luminosities. Supernovas in binary star systems (Type Ia supernovas) have a small range of luminosities and can be used as distance indicators. Binary stars consist of two stars orbiting each other, bound together by their mutual gravity. As the stars evolve, mass can be transferred from one to the other (Fig. 3.7). A white dwarf accretes matter until it reaches the limit of 1.4 solar masses, at which point it explodes. The exact cause of the explosion is not known. Possibly, the star collapses to a neutron star, or the hydrogen in its atmosphere becomes so hot that it ignites and starts to fuse, producing a huge nuclear detonation. The fact that Type Ia supernovas explode when their progenitors all have the same mass explains why their peak luminosities are very similar.

The search for high luminosity distance indicators is a holy grail of astronomy. It was the dream of Hubble and Sandage to measure the curvature of space by measuring the apparent brightness of objects of known luminosity located at increasingly large distances from us. The problem was that objects detectable at large distances, such as quasars, have a wide range of luminosity. If we were studying light bulbs instead of stars and all light bulbs were 75 W light bulbs, we could tell the distance to any light bulb at night by seeing how bright it appeared to our eyes. If however light bulb luminosities (wattages) varied from say 60 to 80 W, a bright light bulb in the sky could be a 60 W light bulb nearby or an 80 W light bulb a bit further away. We want to know the scatter in wattages so to speak of a given class of astronomical object such as quasars or supernovas. The smaller the scatter, the more accurately we can calculate the distance to the object.

In order to measure the curvature of space or density of the universe, one has to be able to measure the apparent brightness of object of known luminosity. Why not use Cepheids, the variable stars that Hubble detected in the Andromeda nebula? The most luminous Cepheids are indeed quite bright, the brightest are about 30,000 times as bright as the Sun. With the Hubble Space Telescope, we can use Cepheids to measure the distances to galaxies as far as 50 million light years from our galaxy. The problem is that to measure the curvature of space we need to measure the distances to objects hundreds of times further away than Cepheids. Supernovas reach peak luminosities of 10 billion solar luminosities, much more luminous than Cepheids. Supernovas are thus a perfect tool for cosmology. We will discuss these findings that were central to the discovery of dark energy in Chap. 7.

Galaxies

We know that galaxies are immense aggregations of stars, gas, and dust and dark matter held together by gravity. A wide variety of

FIG. 3.8 Similar in size to the milky way, the spiral galaxy NGC 3370 lies about 100 million light-years away. The Hubble Space Telescope's Advanced Camera for Surveys has imaged Cepheids stars in this galaxy to accurately determine NGC 3370's distance. NGC 3370 was chosen for this study because in 1994 the spiral galaxy was also home to a well studied stellar explosion – a Type Ia supernova. Combining the known distance to this standard candle supernova, based on the Cepheid measurements, with observations of supernovas at even greater distances, has helped to reveal the expansion rate of the entire Universe itself (Credit: NASA, ESA, Hubble Heritage STScI/AURA)

galaxies appear in images taken with telescopes. Broadly speaking galaxies can be placed in one of three categories: spiral galaxies, elliptical galaxies and irregular galaxies. The visible universe contains over 50 billion galaxies.

Spiral galaxies (Fig. 3.8) contain flat disks with bulges at their centers. The disks contain cool gas and dust interspersed with hotter gas and they exhibit spiral arms. The disks show evidence of ongoing star formation. The disks of spiral galaxies lie within a halo that can extend out to over 100,000 light years. Most galaxies (about 80 %) in the universe are spirals, and they usually have fewer close neighbors than elliptical galaxies.

FIG. 3.9 Elliptical galaxies lack gas and dust to form new stars. Their randomly swarming older stars, give them an ellipsoidal (egg-like) shape. M87 is the dominant galaxy at the center of the Virgo Galaxy Cluster. This elliptical galaxy is over 120,000 light-years in diameter (Credit: Robert Gendler www.robgendlerastropics.com)

Elliptical galaxies (Fig. 3.9) lack a disk component and, indeed, look like footballs in shape. They contain very little gas and dust, although they do contain some very hot gas which emits X-rays. Elliptical galaxies are more common in clusters of galaxies, and they come in a wide range of masses, sizes, and luminosities. The largest known galaxies are elliptical galaxies and are about 50 times the size of our Milky Way.

Our third class of galaxies, irregular galaxies (Fig. 3.10) have a white to bluish color and are less organized than ellipticals or spirals. The nearby Magellanic clouds are an example of irregular galaxies. As we look into the past the fraction of galaxies classified as irregular rises dramatically. Dwarf irregular galaxies contain relatively high levels of gas, and are thought to be similar to the earliest galaxies that populated the universe. Some irregular galaxies are small spiral galaxies that are being distorted by the gravity of a larger neighbor.

FIG. 3.10 NGC 4449 is an irregular galaxy. It is less than 20,000 light-years across, similar in size to one of our Milky Way's satellite galaxies, the Large Magellanic Cloud. This Hubble Space Telescope image highlights the *reddish* glow of hydrogen gas which traces star forming regions. Features include interstellar arcs and bubbles blown by short-lived, massive stars (Credit: Data – Hubble Legacy Archive, ESA, NASA; Processing – Robert Gendler, www. robgendlerastropics.com)

Cosmic Mayhem: Galactic Collisions

The nearest large spiral galaxy to our own is Andromeda, about 2 million light years away, it is visible to the naked eye on dark nights as a faint, fuzzy patch of light. As the universe expands galaxies are drawn away from each other. In some instances two galaxies can be close enough that their gravity overcomes the expansion and the galaxies start to move towards each other. The Andromeda galaxy feels the pull of gravity from our own galaxy and is, in fact, moving toward us at about 300,000 miles per hour. The speed at which the two galaxies are approaching each other is increasing as they get closer to each other. Eventually (in about 2 billion years) the two galaxies will sideswipe each other and eventually merge to form a single elliptical galaxy. It is unlikely that individual stars will actually collide during this galaxy collision, why?

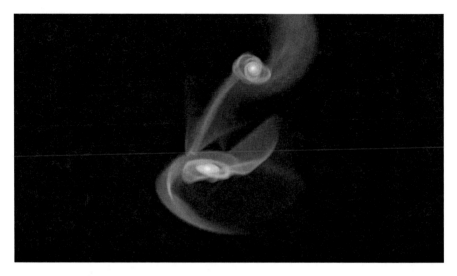

FIG. 3.11 A computer simulation of the future collision of the Milky Way and Andromeda. The simulation used a total of 310 million particles. The simulation shows the mixing of the old (*red*) bulge and young (*blue*) disk stars as the galaxies merge. Simulations can be used to study the details of the merging process and its consequences for the structure in elliptical galaxies. An actual example of such a merger is shown in Fig. 3.12 (Credit: John Dubinski, University of Toronto)

If we were to shrink the Sun to the size of a basketball, the nearest star would be about 3,000 miles away. In contrast, if our galaxy were shrunk to the size of a basketball, Andromeda, the nearest large spiral, would only be about 4 m away. One can infer from these rough estimates that collisions between stars are very, very unlikely. It is equally clear, that since galaxies are separated by only a few times their diameters, galaxy collisions will occur fairly frequently in the universe. The collisions take hundreds of millions of years to occur, so we can, at best, see mere snapshots of galaxy collisions occurring now.

Because galaxies were closer together in the past mergers (collisions) occurred more frequently. Images of galaxies at high redshifts show more distorted and irregular shapes, suggestive of merging events. It is absolutely remarkable that we can take pictures of galaxies as they appeared when the universe was less than a billion years old and compare these with galaxies that we see around us today.

Figure 3.11 shows the results of a galaxy collision in a computer simulation. Depending on the relative masses, separations,

FIG. 3.12 A strongly interacting pair of spiral galaxies. The image displays tidal arms. The bridge between the galaxies is created by tidal forces. Galaxy mergers can trigger high rates of star formation (Credit: NASA, ESA, the Hubble Heritage Collaboration, and A. Evans, University of Virginia, Charlottesville/NRAO/Stony Brook University)

and orientations of the merging spirals, objects with rings and tails can form. Galaxies like these are occasionally seen in nature (Fig. 3.12). Computer models also suggests that two spirals in collision can give rise to an elliptical galaxy. This hypothesis may explain why so many ellipticals are found in dense environments like galaxy clusters.

Mergers are believed to be associated with star formation. If turning spiral galaxies into elliptical galaxies necessitates large bursts of star formation, have these been observed? Possibly. Some galaxies (starburst galaxies) are very bright at infrared wavelengths and show disturbed morphologies. They have extended tails and ring-like structures, which can be explained by merging. Stars form much more frequently in these galaxies than in the Milky Way.

When we think of human development we discuss the role of nature versus nurture. Was a person born with a certain talent, or was that talent nurtured by the environment that person was in? As with human beings both effects play a role in the lives of galaxies. We shall see in Chap. 10 that we can use galaxies as fossils to learn about the distant past, but galaxy collisions make us realize the limitation of this approach.

We have come a long way from the Greek vision of the cosmos. The modern cosmos is a very turbulent place, where stars are constantly being born and are exploding, and where galaxies collide in dramatic events that change their appearance forever.

Supermassive Black Holes: The Monsters at the Center

Galaxies called active galaxies have extremely bright central regions. The emitting region is so small it can look like a bright star in images of the galaxy. Such galaxies usually look like normal elliptical, or spirals, but they radiate enormous amounts of energy at radio wavelengths. These galaxies are known as active galaxies. The central region of these spiral galaxies varies in brightness over a timescale of a few weeks. We can use this fact to set a limit on the size of the emitting region. Since the fastest speed at which information can be conveyed is the speed of light, the emitting region must be smaller than a few light months in size. To understand this let us imagine that you and a bunch of friends are spread out in space and you want to give the signal to everyone to turn on their flashlights. The fastest way to do this is to tell them to turn their flashlights on when they see yours go on. The signal from your flashlight travels at a finite speed, the speed of light. Assuming your friends react immediately if they are located a light year away they will turn their light on 1 year after you. The ones located a light month away will take 1 month to react to your signal. Thus a region of space filling 1 light year will take about 1 year to have all the flashlights go on. This is in essence the argument on the limits of the size a region emitting light based on the time it takes for the light to vary in brightness.

The small sizes derived from the light variability arguments suggest that some novel mechanism must be at work for producing the energy we see. The central light year of an active galaxy produces more light than all the 100 billion stars in the whole galaxy.

Some elliptical galaxies also show strange activity at their centers. Associated with this activity are enormous blobs of radio emission connected to the centers by jets of radio emission. The jets can be up to several million light years in length. These galaxies known as radio galaxies are so powerful that one can see them out to enormously large distances. Indeed, for a while, the most distant objects known were radio galaxies. It is interesting that radio astronomy started when the developers of radar during World War II noticed radar echoes coming from the upper atmosphere which they associated with meteor trails. When the war ended, radio astronomy came into existence as the scientists followed up on these observations. It is a spectacular example of serendipity in research, that a study of radio emissions in the upper atmosphere led to the discovery of the most distant and most luminous known objects in the universe.

When the surrounding galaxy is too faint to detect and we can only see the star-like central light we call the object a quasar. Since they were originally discovered as radio sources, they were named quasi-stellar radio sources. It turned out that these objects were very distant active galaxies.

Active galaxies present us with a problem. We have to understand how so much energy can come out of a small central region. The answer lies in the fact that super massive black holes lie at the centers of some, if not all, galaxies (including our own). Observations of the centers of galaxies suggest that regions less than a light year across may contain a black hole with a mass between several tens of millions up to several billion times the mass of the Sun.

It seems paradoxical that black holes are responsible for the most luminous objects known in the universe. Shouldn't black holes be black? The answer to this question is no, because matter on its way into the black hole is accelerated to tremendous speed and gets very hot. The heating is caused by friction. Since hot matter emits lots of light, the region surrounding the black hole can glow very brightly. Once matter gets close enough to the black hole, nothing, not even light can escape, and the black hole is indeed invisible.

The reason the central region of an active galaxy can produce so much energy is linked to the efficiency of energy production. In the nuclear fusion process less than 1 % of the mass involved in the fusion reaction gets converted to energy. However, calculations suggest that as much as 20 % of the total mass of infalling matter can be converted to energy (i.e. light) before the mass is lost forever in a black hole. It is plausible that a star like the Sun could be swallowed by a central massive black hole. This would release enough energy to power an active galaxy for a 1,000 years. Galaxy mergers discussed above could send a star into the central black hole, thereby turning the galaxy into an active galaxy. We have strong evidence that our own galaxy contains a supermassive black hole, so it is possible that it, too, was once a quasar. Surveys of distant quasars indicate that the quasar phenomenon was more frequent in the past. In other words, a higher fraction of all galaxies were quasars at a redshift of one than today. This is possibly due to the fact that collisions and mergers were more frequent in the past. It is another instance of looking back in time and seeing a universe different in appearance from the one we see today.

What Lies Between the Galaxies?

Our best estimate from Big Bang nuclear fusion calculations and observations of the background radiation is that 4.6 % of the density of the universe consists of baryons in the form of neutrons and protons. But where are these baryons located? We estimate that roughly 10 % of the baryons are seen as visible matter in the form of stars in galaxies. Hot gas in clusters of galaxies account for another 5 %. Quasar spectra suggest that another 30 % is in the form of hydrogen clouds located between the galaxies, the so-called Lyman-alpha forest. It is possible that another 10 % lies in the form of warm intergalactic gas. Nevertheless about 50 % of the baryons remain unaccounted for. One idea is that these "missing baryons" may be difficult to detect because they are concentrated in a filamentary web of tenuous warm gas between galaxies that has been continuously heated during the process of galaxy formation.

The hot gas in clusters illustrates how we detect gas between galaxies. This gas is detected because it is hot enough to emit X-rays. Since X-rays are (fortunately for us) absorbed by the atmosphere, X-ray astronomy must be conducted by satellites orbiting the Earth. The role of telescopes is to collect light from distant objects and bring it to a focal point. For optical telescopes, this is done with lenses or reflecting mirrors. X-rays, because of their high energies, tend to pass straight through or be absorbed by the material they strike. The mirrors used to create X-ray images are called grazing incidence mirrors because they focus light that hits their surface almost parallel. X-ray telescopes have improved tremendously in the last few years, such that we can now measure redshifts using X-ray data.

When the first X-ray images were taken by a satellite known as the *Einstein Observatory*, it became clear that clusters of galaxies contained a lot of hot gas with temperatures of about 10 million degrees Kelvin. A substantial fraction, as much as 30 % of the cluster mass, is believed to consist of this gas. A number of questions arise, including, the obvious: "How did the gas get there?" Our current belief is that some of the gas is a relic from the time when the cluster formed. Some of the gas also must have originated in galaxies, because we can detect emission from iron in the cluster X-ray spectrum. Since no iron was created in the Big Bang, the gas must have been processed inside stars, expelled into interstellar space by supernova explosions, and finally blown out of the galaxy during its journey in the cluster. The second question is "Why is the gas so hot?" It has been found that the temperature of the X-ray gas increases for more massive clusters. There is a rule of thumb that the galaxies in clusters that are more massive tend to move faster than galaxies in less massive clusters. By move I mean speed of motion through space, not rotation speed. We shall discuss this more in the next chapter. Galaxies in clusters move at speeds of about $800 \, \text{km s}^{-1}$. From the theory of gases, we can associate a speed with a temperature; the temperature of the gas tells us how fast the molecules constituting that gas are moving. It turns out that the velocity of the gas molecules in clusters inferred from the temperature is comparable to the velocity of galaxies. The idea is that the gas is stripped from galaxies and then stirred up by turbulent motions until it reaches these high temperatures.

The Visible Universe Across the Electromagnetic Spectrum

Let us review the information we get from visible matter by wavelength of light received. We start with radio waves, which have the longest wavelengths, and proceed through visible light, all the way up to gamma rays.

At radio wavelengths, we detect the light from neutral hydrogen atoms. Spiral galaxies contain hydrogen gas, and we can map the distribution and speed of this gas using radio telescopes. Radio telescopes also detect radiation from electrons spiraling in magnetic fields. We see tremendous amounts of this radiation emitted by radio galaxies, which emit most of their light at radio wavelengths. With radio telescopes, we can also detect remnants of supernova explosions known as pulsars. There are several radio telescopes in various countries. A prominent one open to visitors is the Very Large Array, a spectacular array of 27 radio dishes located near Socorro, New Mexico. These dishes are spread out over a distance of several miles on rails.

At shorter wavelengths, we encounter infrared emission, which is mostly associated with warm clouds of gas and dust surrounding star-forming regions. Infrared astronomy can be carried out from high altitude dry observing sites such as Mauna Kea in Hawaii. You may have visited Hawaii and noticed how humid parts of these islands are. The summit of Mauna Kea is very dry because of the high altitude. Satellites also take infrared telescopes above the Earth's atmosphere. Infrared wavelengths are less strongly scattered than shorter wavelength light, so we can use infrared light to search for objects that would be obscured from view by gas and dust.

Most stars emit most of their energy at visible wavelengths. This is because of their surface temperatures. We can also study gas clouds at visible wavelengths, gathering information about the composition, temperature, and density of the emitted gas. Because the light from distant galaxies gets shifted toward the red, for high redshift galaxies, the light which is emitted as ultraviolet radiation is detected as visible light or even infrared light by the time it reaches the Earth.

Ultraviolet light does not penetrate the Earth's atmosphere, so we study it using telescopes above the atmosphere, such as the Hubble Space Telescope. Ultraviolet light is emitted by young, very massive, hot stars, among others. It is important to know what nearby galaxies look like at ultraviolet wavelengths so that we can compare like with like when we compare nearby galaxies with high-redshift galaxies. Going on to X-rays, we have mentioned emission from the gas in clusters. The central regions of active galaxies also emit X-rays.

At the shortest observable wavelengths we detect gamma ray bursts, brief flashes of gamma-ray energy, lasting from a few milliseconds to a few hundred seconds. These short bursts of gamma radiation were observed to be distributed all over the sky. In 1999, a number of these bursts were identified with galaxies at high redshifts, which suggests that the bursts generate enormous power in very short times. The intrinsic power of the first gamma-ray burst identified with a distant galaxy was estimated to be about 10^{16} (or 10 million billion) times that of our Sun. We think that gamma-ray bursts are supernova events that produced intense narrow jets of radiation. In 2008 a gamma-ray burst went off that was visible to the naked eye for about 30 s. This object is 7.5 billion light years away so that event happened long before the solar system even existed.

Galaxies and Cosmology

We see objects because they shine; that is to say, they emit electro-magnetic radiation. Stars are the most obvious example of visible matter in the universe. They emit light because they produce energy in their centers from nuclear fusion. Stars are assembled into larger systems called galaxies. It is the study of these systems that concerns cosmologists. Most of the light emitted by galaxies is produced by stars. Galaxies also contain gas and dust, which we detect through its absorbing properties. Active galaxies produce substantial amounts of energy from a source other than nuclear fusion. These galaxies are believed to contain very massive black holes at their centers. Matter falling into the black hole is believed to be the source of energy for the radiation emitted by these galaxies. When no matter is falling into the black hole the central region is very faint but we still sense the presence of the black hole from its effect on the orbits of nearby stars.

Cosmology seeks to explain and organize all this information about galaxies. We wish to explain the world of galaxies in terms of the history of their formation. The driving force behind galaxy formation is the gravity of dark matter halos that formed in the early universe. Dark matter is the subject of the next chapter.

Further Reading

Galaxy Collisions. C. Struck, New York, Springer-Praxis, 2011.

Galaxies and the Cosmic Frontier. W. H. Waller and P. W. Hodge, Cambridge, Harvard University Press, 2003.

4. Dark Matter

...the closest I came to an exam was when one day Ehrenfest asked me to recite and discuss Maxwell's equations – the fundamental equations of electricity and magnetism – while walking up and down the corridor. "Yes, you have understood some of the music" was his final verdict.

Hendrik Casimir, Haphazard Reality

When astronomers carry out a census of the universe they find that stars contribute only 1 % to the total density of the universe. Ordinary matter (such as the atoms the make up our bodies) contributes 4.6 % to the total density of the universe. We know that 95 % of the stuff in the universe is invisible to us. Twenty three percent of the density of the universe consists of dark matter; weakly interacting particles that have not been directly detected on Earth. The remaining 72 % consists of dark energy, an unknown force that counteracts the effects of gravity. These conclusions are remarkable for at least two reasons; firstly the numbers themselves and secondly the undreamed of accuracy with which they have been measured. In this chapter we show how these conclusions were reached using telescopes and astronomical satellites as well as presenting the arguments for and against the existence of dark matter.

The Discovery of Neptune: Putting Sir Isaac's Theory to Work

A powerful aspect of Newton's theory of universal gravitation is that one can make predictions about the behavior of matter. This applies, for example, to motion in the solar system. In 1781, William Herschel discovered the first planet to be found with a telescope. Using Newton's theory, an accurate orbit for the new

G. Rhee, *Cosmic Dawn: The Search for the First Stars and Galaxies*,
Astronomers' Universe, DOI 10.1007/978-1-4614-7813-3_4,
© Springer Science+Business Media, LLC 2013

planet, named Uranus, was established. Uranus is four times larger than the Earth and about 20 times farther from the Sun. It takes about 3 h for light to travel from the Sun out to Uranus. This planet has a ring and at least 15 moons. Intriguingly, it is tipped on its side. The axis of rotation of the planet lies in the same plane as its orbit. This is possibly the result of a collision that happened long ago. Herschel proposed naming the planet George after King George III. Unfortunately, in this author's opinion, this name was not adopted. As the planet's position in the sky was measured during the following few decades, it became apparent that Uranus's motion was not consistent with the prediction based on Newton's laws. The difference between the actual and expected position was 20 arc seconds from 1790 to 1830. By 1840, it had increased to about a minute of arc i.e. 60 arc seconds or 1/30 of a moon diameter. I sometimes quote angular measurements in moon diameters, because this is an object we are used to seeing in the sky. Astronomers usually use degrees and minutes and seconds of arc. A minute of arc is 1/60 of a degree. A second of arc is 1/60 of a minute of arc or 1/3,600 of a degree. The Moon, like the Sun has an angular diameter of about half a degree, or 30 min of arc.

When faced with such a discrepancy, astronomers could either dismiss Newton's laws as having been proven wrong or try to reconcile the known facts with Newton's laws. They chose the latter course. One reason for this was that Newton's laws had accurately described the orbit of Halley's Comet, observed in 1758–1759. The reasonable option was thus to postulate the existence of a perturbing object, whose gravity was influencing the path of Uranus around the Sun. A young Englishman, John Adams, and a well-known Frenchman, Urbain Leverrier, predicted the existence of an undiscovered planet based on the motion of Uranus. Adams requested that a certain part of the sky be searched, but he could not convince the astronomical establishment to carry out the project. His work was subsequently recognized, to the extent that the house I lived in while an undergraduate at Cambridge was located on Adams Road. During Adams' third year as an under-graduate, he decided to study "the irregularities of the motion of Uranus... in order to find out whether they may be attributed to the actions of an undiscovered planet beyond it." Such a display of initiative and technique in a 22 year old is really impressive.

In the summer of 1846, Leverrier repeated Adams's calculation and sent a letter suggesting a search for the planet to Johann Galle at the Berlin Observatory. On the night of September 23 1846, Galle pointed his telescope to the position suggested by Leverrier and saw the planet Neptune. It was a triumph for Newton's theory. Further discrepancies were discovered in Uranus' orbit, which prompted a search for yet another planet. Pluto was discovered after an extensive search by Clyde Tombaugh in 1930. I was fortunate to meet him while I was doing postdoctoral research at New Mexico State University in the early 1990s. Arago, another distinguished French scientist, noted of Neptune's discovery that "In the eyes of all impartial men, this discovery will remain one of the most magnificent triumphs of theoretical astronomy."

Leverrier discovered a discrepancy in the motion of Mercury in 1855 and thought this was also due to the existence of an undiscovered planet, which he called Vulcan. In this case, however, it was indeed a shortcoming of Newton's theory that was responsible. Einstein's theory of general relativity provided the correct explanation for the discrepancy.

We tell this story, not because of a sudden need to discuss planets in a cosmology book but because it illustrates the central theme of this chapter. The presence of visible matter can be used to detect the presence of other matter. In the case of Neptune the matter was visible. The presence of an invisible or dark object could just as well have been inferred using this method. As one of Leverrier's colleagues pointed out, "he discovered a planet with the tip of his pen, without any instruments other than the strength of his calculations alone." This method has been applied to the motion of stars and gas within galaxies, to the motions of galaxies within clusters, and, indeed, to the motion of matter in the universe. In each instance we use the motion of visible matter to infer the existence of invisible matter using the known laws of physics.

The method has its pitfalls. As we have mentioned, Leverrier believed he had found a new planet, which was perturbing the motion of Mercury. In this case, it was Newton's laws that were in error. Unfortunately there is no recipe for telling whether we should prefer to believe in the existence of dark matter or the falsity of Newton's laws. Newton's laws have not been tested on the scales on which we observe galaxies and clusters of galaxies.

Most (but not all) astronomers today believe in the existence of dark matter. Martin Rees suggests that "we keep our minds open (or at least ajar) to the possibility that our ideas on gravity need reappraisal."

Dark Matter in Our Own Galaxy

We can apply the method used to discover Neptune to search for dark matter in our own galaxy. It can be shown mathematically that the average speed of a star in a galaxy is related to the strength of the gravitational force that the star is experiencing. The gravitational force is caused by all matter, not just visible matter. We can thus use the strength of the gravitational force inferred from the motion of visible objects to calculate the total mass that is present. This is the theme of this chapter. The whole issue can be formulated as a simple question: *Does there exist in the universe a vast amount of dark matter, which remains undetected except for its gravitational influence?*

The first evidence for dark matter outside the solar system came from the study of the motion of stars in our galaxy in the early twentieth century. Let me begin by giving a quick sketch of the structure of our galaxy, which is believed to be an average spiral galaxy. The Milky Way consists of three major structures; a disk, a bulge, and a halo shown in Fig. 4.1. The disk is a gigantic rotating structure about 100,000 light years in diameter. The disk is a very flattened structure with a thickness of 2,000 light years. Our Sun, which is located in the disk, is located about 30,000 light years from the center of our galaxy. At this distance from the center, the disk rotates at about $200 \, \mathrm{km \, s^{-1}}$. It thus takes about 250 million years for the Sun (and of course the planets) to rotate once about the center of our galaxy. The halo of our galaxy is a spherical structure consisting of about 200 globular clusters. The structures are located within this sphere. Shapley used the globular clusters to estimate the size of our galaxy. Globular clusters are star clusters that contain as many as a million stars in a region 60 light years in diameter. The third component of the Milky Way is the nuclear bulge. The stars in the bulge are similar to halo stars (old stars) whereas the disk is the site of ongoing star formation and contains very young stars as well as older stars like our Sun.

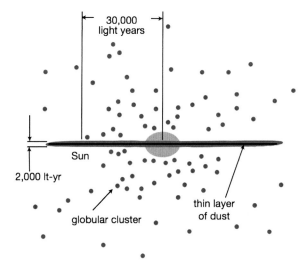

FIG. 4.1 Overview of the structure of the Milky Way galaxy. There are three major components: a disk of stars, gas and dust; a central bulge; and a halo of old stars. The galaxy shown here is embedded in a dark matter halo several times its own size (Fig. 4.4)

As they rotate about the center of the galaxy, disk stars such as our Sun also move vertically, oscillating about the plane of the disk. If we think of the disk as lying in a horizontal plane, the disk stars have a vertical component to their motion. By measuring the vertical speed of these stars and the thickness of the disk, we can estimate the strength of the vertical gravitational field due to the disk near the Sun. The vertical velocities are about $20\,\mathrm{km\,s^{-1}}$, 10 times less than the rotational velocities. It turns out that the method is difficult to implement in practice, but some studies suggest there is between two to five times as much dark matter as visible matter in the solar neighborhood.

Since the Sun rotates about the center of the galaxy, we can use the rotation speed and distance from the galactic center of the Sun to estimate the mass of the Milky Way. This method is a simple application of Newton's laws. We use the same method to estimate the mass of the Sun from the period and size of Earth's orbit, or the mass of Jupiter from the period and size of the orbits of its moons. In this manner, we calculate the mass of our galaxy to be about 100 billion (10^{11}) solar masses.

We can actually do better than this and learn about the *distribution* of mass in our galaxy, not just the total mass. Hydrogen

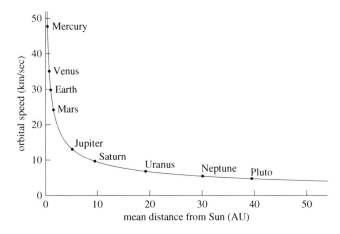

FIG. 4.2 Rotation speed of the planets around the Sun plotted against their distance from the Sun. Since almost all the mass in the solar system is in the Sun the more distant planets rotate slower

atoms give off radio waves having a wavelength of 21 cm. The disk of our galaxy contains substantial amounts of hydrogen gas. With radio telescopes, we can measure the rotation speed of this gas at various distances from the galactic center. We can use this information to plot what we refer to as the rotation curve of our galaxy. We can do the same for our solar system. Instead of using gas we use the planets as tracers. We know the orbital speed of the planets, and we know their distances from the Sun. The orbital speed decreases as we go farther and farther from the Sun, as shown in Fig. 4.2. The Earth orbits the Sun at about $30\,\mathrm{km\,s^{-1}}$. The planet Saturn, 10 times farther from the Sun than the Earth, has a speed of $10\,\mathrm{km\,s^{-1}}$, Pluto, which lies 40 times further from the Sun than the Earth, orbits at about $5\,\mathrm{km\,s^{-1}}$. The reason for the decrease in speed as we move out from the Sun is that the Sun's gravitational force is weaker at the distance of Saturn than at the Earth's distance.

When we plot a rotation curve for the Milky Way, we find a surprising result. As we see from Fig. 4.3, the rotation speed does not decrease with increasing distance from the galactic center. If anything, it increases slightly. The Sun revolves at about $200\,\mathrm{km\,s^{-1}}$ around the center of our galaxy. The rotation speed twice as far out is about $250\,\mathrm{km\,s^{-1}}$. This immediately tells us that the mass in our galaxy is not distributed like the mass in our solar system. In the solar system, the mass is nearly all concentrated in one blob at the center, the Sun. The mass in our galaxy is

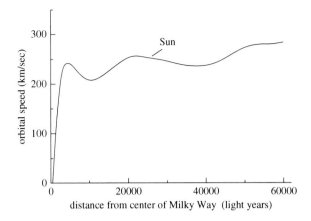

FIG. 4.3 Rotation speed of gas clouds around the Milky Way versus distance from the center of the galaxy. The fact that the shape of the curve is so different to the shape of the curve in Fig. 4.2 tells us that the mass in the Milky Way is much more spread out than the mass in the solar system

much more spread out. A detailed analysis suggests that most of the mass in our galaxy is distributed in the halo of our galaxy. We infer that the halo contains about 10 times as much mass as the disk. However, there is relatively very little light coming from the halo. The globular clusters cannot account for the halo mass. The conclusion is, therefore, that most of the mass of our galaxy is not in the form of visible matter (Fig. 4.4).

Gravitational Lensing and the Search for Dark Matter in the Milky Way

So far, the only two things we know about the dark matter is that it has mass and that it does not emit radiation. We would like to know more. An ingenious experiment has been devised to study this issue. This involves a process known as gravitational lensing, which we encountered in Chap. 1. Gravitational lensing is the bending of light by matter. The idea was proposed in 1916. In 1919, a British expedition observed an eclipse from Brazil and confirmed Einstein's General Theory of Relativity. It has been suggested that the dark matter in our galaxy consists of objects of comparable mass to the planet Jupiter, which never got hot enough in their centers to become stars and start glowing brightly. These objects,

FIG. 4.4 We conclude from using Newton's laws together with the measurements in Fig. 4.3 that our Milky Way galaxy is embedded in a halo of dark matter which contains about 20 times as much mass as that which is visible in the form of gas and stars

brown dwarves, are also known as massive compact halo objects, which gives the acronym MACHO. If a MACHO crosses the line of sight between us and a background star, it will act as a gravitational lens and cause the apparent brightness of the star to increase. These chance alignments are quite rare and should affect about one star in a million each year. Large-scale monitoring projects are under way to detect such lensing events and are finding several such events per year. The lensing objects seem to have a mass between a tenth and half that of the Sun. These mass limits suggest that the MACHOs are white dwarves, dead remnants of stars like the Sun and much less luminous. We therefore have direct evidence for MACHOS, but the current belief is that they are not present in sufficient numbers to account for all the dark matter in the halo of our galaxy. The latest studies reveal that the proportion of the dark halo mass contributed by the MACHOS could be as high as 20 %.

The sheer numbers involved in the MACHO project are impressive. The data are obtained using a 50 in. telescope at the Mt. Stromlo and Siding Spring Observatories in Australia. Since 1992, 27,000 images have been obtained, and the variability of almost 20 million stars has been determined. From this database 50 lensing events have been extracted. Analyzing these huge amounts of data is done automatically using computers. Another project, OGLE, the Optical Gravitational Lensing Experiment, got under way in 1992 as well. More recently, a 1.3 m telescope at the Las Campanas Observatory in Chile has obtained similar data to the MACHO project. These projects involve collaborations between U.S., Polish, and Australian scientists, a good example of successful international collaborations in scientific research. The results of these projects are not yet conclusive as to the origin of the halo dark matter, but they suggest that at least some of the dark halo may be comprised of stellar remnants such as white dwarfs.

Dark Matter in Other Galaxies

Of course, our galaxy is but one of many. It is, in fact, easier to study dark matter in galaxies other than our own. For spiral galaxies whose internal motions are dominated by rotation we have to determine rotation curves. This is done at optical and radio wavelengths. At radio wavelengths, we use radio telescopes to measure the redshifts of hydrogen atoms in the galaxy by observing the 21-cm line emitted by neutral hydrogen. For most galaxies, this gas can be detected farther out than the stars. This is because the gas can be detected more easily than any stars that may be present.

The rotation curves of spiral galaxies resemble that of our own. The example shown in Fig. 4.5 illustrates this, they rise from the center of the galaxy, and reach a maximum value, and remain flat as far out as we can detect them. By a flat rotation curve, I mean that the gas rotates at constant speed around the center of the galaxy, and this speed does not change as one moves away from the galactic center. The implication is that the mass enclosed within a given radius increases as the radius increases. We can, however, see the gas, as we have mentioned, out to considerably larger radii than the stars. This immediately tells us that the stars cannot be responsible for the missing matter. The missing mass must be dark.

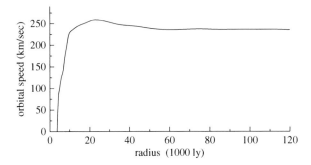

FIG. 4.5 Rotation speed of gas clouds in the spiral galaxy NGC 7332 versus distance from the center of the galaxy. The distance is measured in units of thousands of light years. The measurements thus extend out to 120,000 light years from the center of the galaxy (about twice as far as the Milky Way measurements shown in Fig. 4.3). The plot illustrates that the rotation pattern of our own galaxy is not unique but is common to many other spiral galaxies

The way these observations are usually explained is to assume three components for spiral galaxies: the bulge, the disk, and the halo. The assumption is that the halo contains most of the dark matter. It is also agreed that the density of the halo decreases as one moves away from the center of a galaxy. There are other reasons for wanting spiral galaxies to be embedded in dark halos. Computer models of isolated disks cannot produce stable disks. The disks distort into oval shapes and form large bars. We do not see this in nature. When the models are modified to include the presence of a dark halo, the disks can be shown to be stable. This is of course circumstantial evidence, for no dark halo has been directly detected. We can, however, clearly state that if our theories of gravity are correct, spiral galaxies should contain large amounts of dark matter.

There exists another class of galaxies, ellipticals, that are quite different from spiral galaxies. Elliptical galaxies contain little gas and dust, they do not have disks, and they rotate much slower than spirals. It is much less straightforward to search for evidence of dark matter in ellipticals than in spirals. However, the evidence, once again, seems to point to the existence of dark halos. Elliptical galaxies have been found to contain hot gas at a temperature of about 10 million degrees Kelvin. This temperature is so high that the gas emits large amounts of X-ray radiation, which our telescopes can detect. To confine this gas in a galaxy requires large amounts of dark matter. Because the atoms in the hot gas are moving very quickly, the gravitational field of the galaxy must be strong in order to prevent the atoms from escaping. This in turn requires large amounts of dark matter.

The conclusion is that galaxies seem to be embedded in massive dark halos containing several times the mass of the visible matter and extending to several times the visible radius of the galaxy. This raises the question of the nature of the dark matter. We would also like to know how the dark matter got there. These are questions that theories of the formation of galaxies must try to answer.

Professor Zwicky Weighs the Coma Cluster

Let us now turn our attention to structures larger than galaxies: the great galaxy clusters (Fig. 4.6). Clusters of galaxies can contain many hundreds of galaxies in a region a few million light years in diameter. The Swiss astronomer Fritz Zwicky (1898–1974) was the first to suggest that clusters of galaxies contain dark matter. The method he used is based on a theorem of physics called the "virial theorem". The idea is that for systems in equilibrium, the kinetic energy is comparable to the potential energy. What does all this jargon mean? The kinetic energy of a system is the energy of motion. It is the energy that the system has due to the motion, that is to say, the speed of its constituent parts. For a cluster, the kinetic energy arises due to the speed of the galaxies relative to the cluster center. The potential energy is the energy due to the gravitational force acting between the particles (or galaxies). When we say the system is in equilibrium, we mean that its size does not change with time. Of course, the system is changing because the galaxies are moving, but we mean that, on average, the size of the system does not change.

Clusters of galaxies can be thought of as a swarm of galaxies held together by gravity. We can thus think of each galaxy as a point mass flying back and forth around the cluster. We assume in this discussion that the clusters are in equilibrium. There is evidence to the contrary, but let us proceed with this assumption. We can measure the motion of the galaxies in the cluster by measuring their redshifts. We will notice that not all the galaxies have

FIG. 4.6 The Coma cluster of galaxies. The Coma cluster consists mostly of elliptical galaxies each housing billions of stars (Credit: Dean Rowe)

exactly the same redshift. They have roughly the same redshift, corresponding to the cosmological distance of the cluster, but there is a spread about this average. Some galaxies have a slightly larger redshift than the average, and some have a slightly lower redshift. The spread of redshifts about the mean is a measure of the speed of a galaxy through the cluster. The galaxies in clusters move through the cluster with speeds of about $1,000 \, \text{km s}^{-1}$. We can use this number to calculate the total energy of motion of the cluster galaxies, then using this number we can infer the potential energy using the virial theorem. The potential energy is also a measure of the total mass of the cluster.

To summarize the previous two paragraphs: we can measure the average speed of galaxies in a given cluster and use that to infer the cluster's total mass. We can then add up the masses of all the stars in the galaxies that are cluster members and see if the mass is accounted for. Zwicky, who was ahead of his time in many fields of astronomy did this in the 1930s and found that the sums did not add up. There is a lot more matter in clusters than can be accounted for by the stars in the member galaxies. Clusters contain many galaxies, by the way, up to several thousand for the most massive ones. The conclusion that Zwicky reached in 1933 is that if the galaxies contain stars having a mass like the Sun, then, for every sun, there must be 300 solar masses of dark matter.

Let us take as an example the Coma cluster of galaxies, one of the best studied clusters. The mass contained within 3 million light years of the cluster center is about 6×10^{14} solar masses. Clusters of galaxies are known to emit large amounts of X-rays. This was shown to be the case in the 1960s when the first X-ray telescopes were launched into Earth orbit. The X-rays are believed to be emitted by gas that is extremely hot, as much or more than 100 million degrees Kelvin. We can estimate the mass of the X-ray gas to be about 20 % of the total cluster mass and about 10 times that of the stars in the galaxies. We can also use the temperature of the X-ray gas to estimate the total masses the clustering the same way that we used galaxy motions to estimate the total cluster mass. Let us review the mass budget. Twenty percent of the total cluster mass consists of X-ray emitting hot gas. An additional 2 % of the total mass comes in the form of stars in galaxies. We have thus accounted for 22 % of the total cluster mass, which leaves the vast majority, 78 % of the mass in the form of some undetected dark matter, whose nature is unknown to us. The total cluster masses calculated by the X-ray method and the galaxy motion method agree to a few percent.

There is yet another fascinating way to measure cluster masses: the gravitational lensing method. Clusters are so massive that they can bend the light coming from even more distant galaxies behind the cluster. These background galaxies thus appear distorted in the form of arcs. They also appear bluer and fainter than the cluster galaxies, and they can therefore be distinguished from true cluster members. The shape and extent of the lensing distortions can be used to estimate the cluster mass. The gravitational lensing mass estimates agree with the two previous mass estimates for individual clusters. Since three independent measures provide the same answer, the suggestion is that they are measuring the same thing and that there is indeed a vast amount of dark matter in clusters.

The Bullet cluster of galaxies shown in Fig. 4.7 illustrates the value of comparing different observations of the same object. The

FIG. **4.7** A galaxy cluster commonly known as the Bullet Cluster. The X-ray emission from this cluster is shown in *red* superposed on an image of the cluster taken with the Magellan Telescope in Arizona. The individual galaxies are clearly visible in the optical image. The *blue* hues show the distribution of the total mass of the cluster. This was mapped using gravitational lensing of background galaxies. The fact that the total mass of the cluster is dominated by an invisible component separate from the gas clouds is strong evidence for the existence of dark matter (Credit: X-ray: NASA/CXC/CfA/ M.Markevitch et al.; Lensing Map: NASA/STScI; ESO WFI; Magellan/U.Arizona/ D.Clowe et al. Optical: NASA/STScI; Magellan/U.Arizona/D.Clowe et al.)

fact that the total mass inferred from lensing studies (blue hue) follows the visible galaxy distribution is a persuasive argument for the existence of dark matter. If one was to argue that there is no dark matter, then one would expect the lensing method to reveal a mass distribution that closely follows the X-ray gas distribution; in other words the blue and red images would overlap since the X-rays contain most of the visible matter. The fact that this is not the case and the X-ray clouds appear distorted suggests that we are observing two clusters that have passed through each other (collided) leaving the X-ray gas trailing behind the galaxies and dark matter, as one would expect from simple physical models of two colliding clusters.

Dark Matter Everywhere? The Density of the Universe

We have discussed evidence that the universe is expanding. It is natural to speculate about the future of our universe. The universe will either keep expanding forever or recollapse into a Big Crunch in the distant future. The fate of the universe depends on whether its present density exceeds a so-called critical density. If the universe has a density greater than the critical density it will collapse at some point in the future. Future observers will see this as a change from redshift to blueshift for the light we receive from galaxies. The critical density has a value of 5 atoms per cubic meter, a number determined by the competition between the expansion energy of the universe and the force of gravity. The number Ω measures the ratio of the true density to the critical density. If this number is greater than 1, the universe will recollapse. Let me make clear the Ω is a ratio it is not itself a density.

We might be tempted to say that, of course, the universe has a higher density than the critical density. After all, 5 atoms per cubic meter is much less than the best vacuum achievable on Earth. We must keep in mind, however, that we are talking of the density of the whole universe. We must average out the density of the Earth and the solar system and, indeed, our whole galaxy over vast regions of space. To get a fair measure of the density of the universe, we must sample a large region of space. I live in Boulder City, Nevada. The population density of Boulder City, 1,000 people per square mile, does not reflect the population density of the state of Nevada, which is 50 times less than that.

How are we to estimate the density of the universe? There are a number of different methods, all of which have rather large errors. By combining these methods we can estimate the density of the universe. We can get a lower limit on the density of the universe by considering a volume of space and adding up the masses of all the galaxies within it. The density is then the mass divided by the volume. When we add up the galaxy masses, we must, of course, include the dark galaxy halos mentioned earlier in this chapter. This method yields a value for Ω of about 0.1; that is to say, the density of the universe is 10 times less than the critical density.

We can apply a similar method to clusters of galaxies. We can calculate the masses of clusters and count the number of clusters in a given volume to obtain the density of matter in that volume. This method yields values of Ω of about 0.2. We shall discuss in a later chapter the surveys that astronomers have carried out to map the universe. These reveal that there are large regions of space containing an excess density of galaxies. These regions exert a gravitational pull on surrounding galaxies causing them to fall toward these regions. The motions the galaxies acquire are called peculiar velocities. By studying these peculiar velocities we can estimate Ω. Such studies favor large values of Ω, close to 1.

We can also measure Ω more directly by measuring the curvature of space. General relativity predicts some very strange observational effects. If we take a ruler and observe it at increasingly large distances from us, it will appear smaller and smaller. In other words, the angular size of an object gets smaller at larger distances. This is not true in cosmology however. The principle holds true at small distances, but when we get to very high redshifts of 2 and greater, objects actually appear to get bigger as the redshift increases. This is due to the way light travels through space in an expanding universe. The importance of this effect increases as the density increases. Such an effect could be used to measure Ω. The problem is that we do not have a set of rulers placed at regular intervals in space. We only have galaxies, and they come in many different shapes and sizes. We do not know the intrinsic sizes of objects at cosmological distances from Earth.

The apparent brightness of objects of known luminosity should also depend on Ω. Attempts to measure the curvature of space using objects of known luminosity had been unsuccessful until a decade ago when scientists used supernovae, exploding stars, to estimate the curvature of space. The fact that supernovae occurring in binary systems all have the same luminosity suggests that they can be used as distance indicators to measure the curvature of space and hence the density of the universe. These supernovae act as light bulbs of known wattage distributed throughout space.

Supernovae are fairly rare events. They are believed to go off at a rate of about one per 300 years in our galaxy. The technological challenge that astronomers face is thus to detect supernovae at sufficiently large redshifts and in sufficiently large numbers to measure the curvature of space. Two teams of scientist have made use of large-format imaging cameras and sophisticated detection software to carry out this ambitious project. Current results suggest $\Omega = 1$, a flat universe and furthermore that there is a mysterious force at work in the universe that is countering the force of gravity and causing the expansion of the universe to speed up. This expansion force is associated with an energy known as dark energy which contributes to the energy density of the universe (Ω_Λ).

For over 50 years, cosmologists have dreamed of measuring the curvature of space. We must always be cautious and critical when faced with a spectacular result of this sort. We must ask ourselves what could have gone wrong, or what more mundane interpretation could account for this effect. The observations effectively state that the most distant supernovas are slightly fainter than we would expect them to be in a flat (non-curved) universe. Is there another explanation for this observation? Absorption by dust is a possibility. These issues are currently being debated in the astronomical literature. The current consensus, the result of detailed studies, is that dust cannot be responsible.

The most accurate results are obtained by combining different scientific studies. We discuss the microwave background in some detail in the following chapter. The results of studies of the background radiation using a NASA satellite, the Wilkinson Microwave Anisotropy Probe have been combined with the supernova observations to reach the remarkable conclusion that the universe has a density of $\Omega = 1.00 \pm 0.01$. What is remarkable about this conclusion is not only the value of Ω but also the high precision with which this has been determined. Astronomers can now measure a number that a decade ago was known only to within a factor of ten, to a precision of about 1 %.

What Could the Dark Matter Consist Of?

The supernova measurements combined with the background radiation measurements can not only constrain the total density of the universe but also its different components. The value of the dark energy density responsible for the acceleration of the expansion is believed to be $\Omega_\Lambda = 0.72$ which means that the matter density must be about $\Omega_{matter} = 0.28$ (Fig. 4.8). What could this matter consist of?

We can proceed by ruling out one candidate, the baryons. Baryons are essentially protons and neutrons, the ordinary matter that we are made of and that stars consist of. As we go back in time to the beginning of the Big Bang, the universe gets denser and denser and hotter and hotter. At the earliest times, not even atomic nuclei could have survived the intense heat. When the temperature dropped to a mere one billion Kelvin, atomic nuclei could hold together, and a process called nucleosynthesis occurred. The formation of heavier elements, such as helium and lithium, requires the formation of deuterium, an isotope of hydrogen. The deuterium nucleus consists of a proton and a neutron held together by the strong nuclear force. Initially, any deuterium nuclei that form were blasted apart by the intense heat, but when the universe cooled sufficiently, to a billion Kelvin, the deuterium nuclei could hold together.

Let us turn our attention to the neutrons. A free neutron has a lifetime of about 15 min and will decay into a proton if left on its own. The number of free neutrons left at the time of nucleosynthesis determines the helium abundance. Along with hydrogen and helium, some other light nuclei were produced in the early universe, lithium and beryllium being of particular interest.

Once it forms, the deuterium can then react with more protons and neutrons to form helium. It is in this way that most of the helium that we see in the universe was formed. The abundance of helium relative to hydrogen is not very sensitive to the density of the universe, it is thus a robust prediction of Big Bang theory. In other words it is difficult to imagine a universe in which little or no helium was formed early on. This result is very sensitive to the strength of the strong force that binds nuclei together. If the force were slightly stronger, all the hydrogen would have been converted to helium in the first 3 min. If the force were slightly weaker, no helium at all would have formed.

Not all of the deuterium formed in the early universe goes to form helium. There is a little bit of deuterium left over. The abundance of deuterium is a sensitive measure of the density of baryons in the universe. In a denser universe, the reactions proceed more quickly and less deuterium will survive. It is difficult to measure the deuterium abundance because it is easily destroyed in

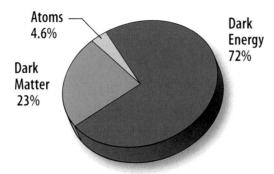

FIG. 4.8 Relative contribution of dark matter, dark energy and atoms to the density of the universe. Ordinary atoms that we are made of contribute a relatively small amount to the total density. It turns out that about 1 % of the density of the universe is in the form of visible stars (Credit: NASA/WMAP Science Team)

stars. Astronomers have searched for deuterium in the atmosphere of Jupiter, in the gas between the stars, and in distant clouds of gas that absorb the light from distant quasars. They find the deuterium abundance to be very low. The baryon density (Ω_b) implied by the measured deuterium abundance is $\Omega_b \sim 0.04$. The measured abundance supports this conclusion. Finally detailed studies of the microwave background variations in brightness from the WMAP satellite conclude that $\Omega_b = 0.046 \pm 0.001$, that is, ordinary atoms make up only 4.6 % of the universe. Again a remarkable conclusion also remarkable for its accuracy, the previous statement is accurate to 0.1 % (Fig. 4.8).

The baryon density today is thus about 1/25 of the critical density. However, we know that the total matter density of the universe is $\Omega = 0.28$. The implication is that most of the dark matter in the universe is non-baryonic. The dark matter cannot consist of protons and neutrons. It must be some other form of exotic matter.

What If the Dark Matter Consisted of Neutrinos?

There are three lines of approach to deciding on the dominant form of dark matter. First, we can try to detect the matter through microlensing experiments. Second, we can try to detect the particles directly and attempt to measure their masses, as has been done for neutrinos. The third way is to study galaxy formation from a theoretical viewpoint and determine which form of dark matter produces galaxies that most closely resemble the ones we see today. The question of the nature of dark matter is still very much open. The most certain thing we can say (and even then, not with total certainty) is that the dark matter exists.

One good candidate particle for dark matter is the neutrino. Neutrinos were produced in large numbers in the early universe with a density of several hundred per cubic centimeter. Given the high density of neutrinos in space, it does not take much mass for a neutrino to play a big role in cosmology. If the mass of the neutrino was about 1/20,000 of the mass of the electron, the neutrino would be the dominant form of mass in the universe. There have been several experiments to directly measure neutrino mass in the last 30 years, and none have produced convincing results. The question of whether the neutrino has mass is nevertheless resolved even though the uncertainties as to what that mass is are still very large.

Swimming Pools Underground: Neutrino Astronomy

There is an indirect way to prove that the neutrino has mass. There are three types of neutrinos known: the tau neutrino, the electron neutrino, and the muon neutrino. The electron, muon, and tau particles are known as leptons, particles that feel the weak interaction. If neutrinos have mass then it is possible that neutrinos could change from one kind to another. Experiments have been carried out in which a beam of muon neutrinos is produced by colliding muons with atomic nuclei. If tau leptons are detected at the point where the muons collide with the nuclei, we have evidence that muon neutrinos have changed into tau neutrinos, since only a tau neutrino can produce a tau lepton by colliding with a nucleus. There is another type of experiment currently under way that studies the particles produced by cosmic rays. The experimental setup can be thought of as a neutrino telescope. Neutrino telescopes are very different from ordinary telescopes.

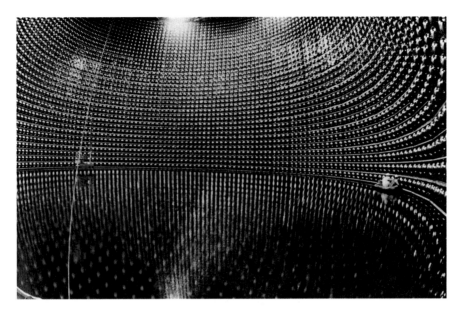

FIG. 4.9 The Super-Kamiokande neutrino detector. It consists of a stainless steel tank 39 m in diameter and 42 m tall filled with 50,000 tons of water. The detector is located 1 km under- ground in the Kamioka mine in Japan (Credit: Kamioka Observatory, ICRR (Institute for Cosmic Ray Research), The University of Tokyo)

The neutrino telescope known as Super-Kamiokande, consists of a tank of pure water installed in a deep mine inside a mountain (Fig. 4.9). As the neutrinos travel through the water, they emit very weak flashes of light that are detected. The detector is direction sensitive, it can tell roughly where the neutrinos came from. The evidence is that more neutrinos are hitting the detector from above than from below. This might seem obvious, but neutrinos interact so weakly that they can travel through the whole Earth quite easily. The fact more neutrinos come from above is interpreted as meaning that some of the neutrinos coming from below are changing from one kind to another on their journey through the Earth and is thus taken as evidence of neutrino mass.

Neutrinos are produced by fusion reactions in the Sun. Based on the Sun's luminosity and our understanding of nuclear fusion reactions, we can predict the rate at which neutrinos should reach the Earth. The first attempt to detect these solar neutrinos took place in the 1960s in the Homestake Mine in South Dakota. This neutrino telescope consisted of a 400,000-liter vat of cleaning fluid! The detector made use of the fact that a chlorine atom can react with a neutrino to produce an argon atom. This reaction is extremely rare–so rare that it is expected to take place only for one atom in the whole tank each day; this in spite of the fact that trillions of neutrinos pass through the tank each second. Neutrinos are very weakly interacting particles. What was intriguing and disturbing was that the Homestake experiment produced only one third as many neutrinos as was expected.

In such a situation where a discrepancy between theory and observation arises, we have several choices. We could blame the experimentalists and claim that the measurement is incorrect. We could also claim that the theoretical predictions are wrong; in other words, we don't really understand the inner workings of the Sun. One way to keep both theorists and experimenters happy is to challenge neither. We can argue instead that both the prediction and the measurement are correct, but that the solar neutrinos (electron neutrinos) are turning into tau and muon neutrinos on the way from the Sun to the Earth. Since the Homestake experiment of the 1960s, three experiments have come online that confirm the original experimental results. The problem, however, is not quite as bad. Instead of only 30 % of the predicted neutrinos being detected 70 % of the neutrinos are now being observed. Maybe we should wait another 30 years to find the remaining third of the missing neutrinos!

The current trend is to argue in favor of neutrinos transforming from one kind to another and to argue that this is evidence for neutrino mass. The limits on the mass of the neutrino are not reliable, but it is still possible that neutrinos could account for a substantial amount of the dark matter in the universe. Neutrinos are also thought to be emitted in very large amounts during supernova explosions. When a supernova was detected in a companion galaxy to ours in 1987, two neutrino detectors were operational and detected a burst of neutrinos coming from that direction. This was a remarkable confirmation of astrophysical theories of supernova explosions. The supernova was visible only in the southern hemisphere, yet the neutrino detectors were in the northern hemisphere so the neutrinos passed through most of the Earth before being detected.

Neutrinos are a form of dark matter known as hot dark matter. They move close to the speed of light in the early universe. This has consequences for galaxy formation which we will discuss in a later chapter. A form of matter that has proven very useful in constructing models of galaxy formation is known as cold dark matter. These particles move around slowly in the early universe. Although they are very popular with theoretical astrophysicists, they have probably not been directly detected on Earth. There are experiments deep underground that are designed to detect these cold dark matter candidates. The search for dark matter particles highlights a theme of cosmology; the link between the very large and the very small. We will see that the properties of elementary particles determine the nature of the largest structures in the universe.

What If Dark Matter Is a Myth?

The astrophysicist Martin Rees has pointed out that "we should leave our minds open, or at least ajar, to concepts that now seem on the wilder shores of speculative". I have presented the conventional arguments in favor of dark matter, most of which I find reasonably convincing. It does no harm, however, to look at unconventional arguments against dark matter and see if they stand up to scrutiny.

The dark matter issue has been with astronomers for over 50 years in the context of galaxies. One of the first discussions concerned the dark matter in the disk of our galaxy close to the Sun. For a very long time that evidence suggested that about half the matter in the disk in the solar neighborhood is dark. It is only quite recently that researchers have argued that there is no dark matter in the solar neighborhood. It is thus possible for age-old evidence to be reevaluated and new conclusions to be reached.

We began this chapter by discussing the motion of the planets, in particular, Uranus and Mercury. The anomalous motions of these two planets were used to predict the existence of two new planets. From Uranus' motion, the existence of the planet Neptune was predicted. From the anomalous motion of Mercury, the existence of the planet Vulcan was predicted. You have probably heard of Neptune. It is a planet that is about 30 times farther from the Sun than the Earth is. It is about 17 times as massive as the Earth and 4 times bigger. I do not imagine that you know many facts about the planet Vulcan since it doesn't exist. The anomaly in Mercury's motion we now know is not due to the influence of another planet. The motion of Mercury cannot, in fact, be entirely explained by Newton's theory of gravity. Einstein's theory of General Relativity can account for all the known properties of Mercury's orbit.

What do these planetary considerations have to do with dark matter and cosmology? The point is that we infer the existence of dark matter by applying Newton's theory of gravitation. As we have seen in the case of the rotation of galaxies, the fact that galaxies have flat rotation curves enables us to infer the existence of dark matter. We believe Newton's laws are a valid approximation of general relativity on these scales and densities. We should keep in mind that this hypothesis has not been tested. It is in fact possible that we are misguided in applying Newton's laws on these scales. Some scientists have given thought to the manner in which Newton's laws would have to be modified to reconcile the rotation speeds of galaxies with the presence of only the luminous matter. These scientists refer to their theory as modified Newtonian gravity. In Newton's theory, the force of gravity falls off as an inverse square law, with distance. If we modify the inverse square law such that, on very large scales. the falloff is slightly slower than an inverse square law we can do away with dark matter in galaxies. The inverse square law has been tested to very high accuracy in the solar system, so we must modify it only on the scales of tens of thousands of light years.

As a temporary advocate of the case against dark matter, I discuss the rotation of galaxies because these provide the most direct evidence in favor of dark matter. The observations of the speed of motion of clouds of neutral hydrogen using radio telescopes are accepted by all astronomers. The observational evidence is very reliable. The issue comes with interpretation. Anyone arguing against dark matter has to provide a viable alternative gravitational theory. It boils down, in a sense, to a matter of taste. Do you prefer to believe in the existence of an undetected component of matter, whose origin and nature is unknown and possibly unknowable, or do you prefer to make a slight modification to a well-tested theory of gravitation? Almost all astronomers choose the former option.

As we turn to larger scales, we reach the scales of galaxy clusters. As we have seen, the same problem arises. Individual galaxies that are members of clusters move too quickly. That is to say, the cluster mass that is inferred from the speeds of the galaxies is larger than the sum of the masses of the galaxies making up the clusters. Is there any way we can challenge this line of argument? The argument is based on applying the virial theorem. This theorem is derived on the assumption that the objects being considered are in equilibrium, that is, the objects are in a stable configuration and are unchanging with time. What should an object in equilibrium look like? An elliptical galaxy provides a good description. One has an elongated mass distribution that is denser at its center than on the outskirts, the density falls off smoothly from the center to the edge of the galaxy. Clusters of galaxies are similar in some ways. They are denser at their centers, and the density falls off as one moves away from the cluster center. Many clusters, however, do show features that are not consistent with the idea that clusters are in equilibrium. Many clusters have a clumpy galaxy distribution. The density distribution shows several peaks, not just one peak at the central location. One can thus make a case that clusters are in fact not in equilibrium, but are young structures that are still forming and one cannot therefore naively apply the virial theorem to these objects. The X-ray maps of clusters support this claim. They also show a clumpy mass distribution with evidence of recent collisions and infall of material. A compromise would be to say that there are a few clusters that do look in equilibrium, and one can reliably estimate the masses of these few clusters. In addition, it seems that estimates of cluster masses based on X-ray, galaxy and lensing observations agree. This may or may not be evidence that the mass estimates are reliable.

To conclude there is a small possibility that there is no dark matter. The strongest evidence in favor of dark matter comes from the rotation of spiral galaxies, and no theory has convincingly shown that this interpretation is false. Let us now review the evidence in favor of dark matter one last time, using a new and important concept the ratio of mass to light.

The Ratio of Mass to Light

We can discuss the amount of dark matter to visible matter in, say, a galaxy by comparing the densities of the dark and visible matter component. A common way of restating the problem is to use mass-to-light ratios. This number is commonly listed in terms of the mass and luminosity of the Sun. In these units, which we shall use from now on, the mass to light ratio of the Sun is 1 (in more conventional units it is actually one half. The mass of the Sun is about 2×10^{33} grams and the Sun's luminosity is 4×10^{33} ergs per second. $2 \times 10^{33} / 4 \times 10^{33} = 0.5$). We can calculate the mass-to-light ratios of stars as a function of their mass. A low mass star having a mass ten times smaller than the Sun has a mass to light ratio of 4,000. In other words, a star that has one tenth of the mass of the Sun is 40,000 times less luminous than the Sun! A star ten times as luminous as the Sun has a mass to light ratio 100th that of the Sun. Such a star is 1,000 times more luminous than the Sun. For the mix of stars close to the Sun in our galaxy, we get a mean mass-to-light ratio of about 2. That is to say if we sum the masses of all the stars close to the sun and divide that number by their total luminosities we get in solar units a number roughly equal to 2. Interestingly, the mix of nearby stars is such that most of the mass comes from stars having less than half the mass of the Sun, while most of the light comes from stars having 1.5 times the mass of the Sun.

What are the mass-to-light ratios of galaxies? If we look at the cores of elliptical galaxies, we get a mass-to-light ratio of about 10. For spiral galaxies, the issue is difficult. The spiral galaxies have flat rotation curves as far out from the center as we can measure. We do not know where the edges of the dark matter halos surrounding spiral galaxies lie. Mass-to-light ratios calculated for spiral galaxies range from about 10 to 200. In terms of their contributions to the density of the universe, the visible parts of galaxies contribute $\Omega \sim 0.01$. If we include the dark matter components of galaxies, we get at most $\Omega = 0.1$, it follows that most of the dark matter is distributed between galaxies.

One way to study the matter between galaxies is to study groups and clusters of galaxies, as we have discussed above. By studying the motion of the galaxies in our local group of galaxies, which includes the Andromeda nebula, a large spiral galaxy, we arrive at a mass-to-light ratio of 100 for the local group. For rich clusters, as we have seen, mass-to-light ratios can also be computed. In the cases of rich clusters containing hundreds of galaxies within a few million light years, the mass-to-light ratio is about 300. As we go to increasingly large objects, we obtain larger and larger mass-to-light ratios. There is more and more dark matter compared to visible matter as we go to larger scales.

We have seen that low-mass stars have mass-to-light ratios that are very high. Why not account for the mass-to-light ratios of 2–300 by choosing a population of stars with suitably low mass? There are two problems here. First, we have to account for the colors of stellar populations. When we look at a galaxy and study its colors, only a certain range of masses is allowed for its stellar population. We cannot arbitrarily decree what masses the stars in a galaxy should have simply to satisfy a mass to light ratio requirement. Second, as we have seen for spiral galaxies, there is a large amount of matter in the outer parts and no visible matter. The halos of spiral galaxies thus *cannot* be made of stars.

We have presented in this chapter a view of the current situation with regard to dark matter. The key facts are that we live in a universe that will expand forever whose density is dominated by dark energy. The stars that shine in the universe only contribute 1 % of the density of the universe. About one quarter of the density of the universe is comprised of dark matter whose nature is at present unknown. We can speculate about the precise properties of the dark matter that are required for galaxies to form and be distributed in space in the manner that we observe. We have yet to detect such a particle using detectors on Earth but there are hints that we may have done so.

Since the formulation of the dark matter problem in the 1930s we have made enormous strides in this field. We have conclusive evidence for the existence of dark energy. We can calculate the composition and density of the universe to an accuracy of about 1 %. Our models of galaxy formation based on dark matter are providing good fits to the data. All this suggests that we are on the right path to unveiling the formation of the first objects in the universe using the Big Bang theory and structure formation driven by dark matter. Some of these points are summarized in Table 4.1.

TABLE **4.1** The composition of the universe

Type	Composition	Main evidence	Amount %
Visible matter	Atoms in stars	Telescopes	1
Baryonic matter	All atoms	WMAP[b]	4.6
Non-baryonic matter	WIMPS[a]	WMAP	23
Dark energy	Cosmological constant	SN & WMAP[c]	72

[a]Weakly interacting massive particles
[b]The NASA satellite Wilkinson Microwave Anisotropy Probe
[c]Supernova observations used in conjunction with WMAP data

Further Reading

In Search of Dark Matter. K. Freeman and G. McNamara. Springer-Praxis, 2006.

The Extravagant Universe: Exploding Stars, Dark Energy, and the Accelerating Cosmos. R Kirshner, Princeton University Press, 2002.

Rotation Curves of Spiral Galaxies. Y. Sofue and V. Rubin in Annual Review of Astronomy and Astrophysics, Volume 39, Page 137–174, Sep 2001.

Part II
The Emergence of Galaxies

Part I of this book described the big bang theory, dark matter and the universe of galaxies and stars that we observe with our telescopes. Part II describes how we think these galaxies and even larger structures emerged in the aftermath of the big bang. We begin in Chap. 5 with a description of the largest known structures in the universe loosely known as the cosmic web. Chapter 6 discusses the physical models we use to explain the formation of galaxies several billion years ago. We are close to predicting which kinds of objects first lit up the universe, and when.

It is natural to see this field of study as speculative but astronomers can measure some of the key numbers the we use in our models with surprising accuracy. In Chap. 7 we show how precision measurements of primordial radiation left over from the big bang have made it possible to determine the age and density of the universe. The cosmic background observations in conjunction with other data have revealed that most of the matter in the universe is invisible (i.e.,dark) and that the expansion of the universe is accelerating in an unexpected and unexplained way.

5. A Map of the Universe

Like the fifteenth century navigators, astronomers today are embarked on voyages of exploration, charting unknown regions. The aim of this adventure is to bring back not gold or spices or silks but something far more valuable: a map of the universe that will tell of its origin, its texture, and its fate.

Robert Kirshner

The smallest object detected with a telescope is an asteroid about the size of a beach ball. The largest asteroids, debris left over from the formation of the planets range are a few hundred miles in diameter. Our sun is almost 1 million miles in diameter. Some red giant stars are large enough that one could fit all the planets up to the orbit of Mars inside them. The largest star we know of is so big that if it were located in the center of the solar system its outer envelope would extend past Jupiter. The Milky Way galaxy disk which we inhabit has a diameter of about 100,000 light years. The largest galaxy we know of lies at the center of a cluster of galaxies and has a huge diameter of 5 million light years.

We can study the universe on such large scales that galaxies themselves appear as dots, just like large cities appear as dots on a globe of the Earth. In the past two decades a large effort has gone into mapping out the structures that are traced by galaxies. The largest known structure in the universe is called the Sloan Great Wall. It was discovered in 2003. The Sloan Great Wall is about 1.4 billion light years in length and is located about 1 billion light years from earth. The structure spans about one quarter of the sky.

How do we discover such gigantic structures? How did they come into existence?

G. Rhee, *Cosmic Dawn: The Search for the First Stars and Galaxies*, Astronomers' Universe, DOI 10.1007/978-1-4614-7813-3_5, © Springer Science+Business Media, LLC 2013

The Large Scale Structure of the Universe

The largest structures in the universe are found by making galaxy maps. We measure the distances to lots of galaxies and plot them on a map. A map of the town you live in will be two dimensional. That is the map can be drawn on a sheet of paper. Our maps of the universe are three dimensional, we use three numbers to specify a galaxy's location in space. The first two are angular coordinates which specify where in the sky a galaxy is located. We could just say a few degrees to the north of the two bright stars in the big dipper. We achieve greater precision by using two numbers to specify where in the sky a galaxy is located. A third number we use is the redshift. The redshift gives us an estimate of the distance to a galaxy.

The first step in planning a survey is to select the galaxies whose redshifts we are going to measure. To understand the biases in our survey we must select galaxies using quantitative criteria. We cannot measure distances to all the galaxies within a fixed volume of space because some galaxies are too faint. We thus have to select some fraction of the galaxies in the volume of interest. The telescope we use can measure redshifts of galaxies to a brightness limit in a reasonable amount of time. So if we specify the condition that a redshift must be measurable in less than 1 h (for example) that in turn specifies the faintest galaxies that we can survey.

To implement this approach we must first image the sky to construct a catalog of galaxies. From this catalog we select a sample galaxies whose redshifts are to be determined. When a survey is completed astronomers will analyze the results very carefully looking for correlations between say a galaxy's color and the color of its neighbor. One must be sure that any effect that is found is not a consequence of the method of observing but rather a reflection of a genuine trend found in nature.

Pollsters face similar problems. Let us say that the pollsters take a poll in a city to find out who will be elected mayor in a forthcoming election. The results show that the republican candidate will win by a large majority. When election time comes the democrat wins by a landslide. What went wrong? It turns out the pollsters only went to the most expensive neighborhoods to take their poll and thus came away with a biased sample. In the jargon of astronomy there was a strong selection effect in the polling. By polling people in affluent neighborhoods the pollsters biased the outcome of the poll. They did not have a fair sample of the population of the town. This is what astronomers must always ask themselves when analyzing data, what man-made error or measurement error could have produced this result. Pollsters face an even greater problem, the polling results when made public can influence the outcome of the election.

Once a galaxy survey has been carried out, the results are used for statistical studies by astronomers. The simplest measurements we can make are of the clustering properties of galaxies. Are the galaxies distributed randomly in space? We already know the answer to that! Does a large structure dominate the survey?

Edwin Hubble's original studies of galaxies included a few dozen galaxies. The analyses have become more sophisticated as the numbers of galaxies surveyed have increased. By 1976 the redshifts of about 2,700 galaxies were known. Surveys by a team from the Harvard-Smithsonian Center for Astrophysics provided strong evidence for the filamentary nature of the distribution of galaxies (Fig. 5.4). The extended survey included about 15,000 galaxy redshifts. The next step involved two large projects. The 2-degree field Galaxy Redshift Survey was carried out between 1997 and 2002 on the 4 m Anglo-Australian telescope in Australia. The survey obtained redshifts for over 200,000 galaxies located within 1,500 million light years of our galaxy. Located at Apache Point Observatory in New Mexico, the Sloan Digital Sky Survey imaged a quarter of the sky and obtained spectra for about one million objects. These projects and the results they delivered will be discussed in more detail below.

We have to do systematic work to make progress in astronomy. Tycho showed us this with his work on the island of Hveen hundreds of years ago. By making careful and very accurate measurements of the planets, Tycho enabled Kepler to determine the true shape of planetary orbits. Galaxy surveys have also revolutionized our view of the universe.

The Redshift Machines

To map the galaxy distribution in three dimensions we need to measure redshifts. Redshifts are measured by obtaining a spectrum of a galaxy using a spectrograph. The spectrograph uses a device such as a prism or a diffraction grating to split the light into its constituent colors. Isaac Newton used a prism to show that sunlight contained all the colors of the rainbow. With a prism and our naked eye we can see what colors are present in the light we examine. The spectrograph enables us to plot the intensity of light as a function of wavelength. It tells us how much light is present at red wavelengths compared to say green wavelengths. The output produced by putting the light through a spectrograph is called a spectrum. By analyzing the spectra of galaxies we can measure their redshifts and hence estimate their distances.

The spectra of galaxies at visible wavelengths consist of the spectra of lots of stars added together. Stellar spectra are known to contain absorption lines, narrow parts of the spectrum where the light is fainter. These can be easily seen in our sun's spectrum. The absorption lines are due to the presence of atoms such as calcium and iron in the outer parts of stars. These absorption lines have known wavelengths that we can measure in the laboratory. Their location in a galaxy's spectrum enables us to measure the redshift of that galaxy. Essentially we see a recognizable pattern of lines all shifted towards the red (Fig. 5.1). Computer programs have been developed to automatically measure redshifts.

The most common spectrograph design was for a long time the long slit spectrograph. Any object that fell on the slit would have its spectrum taken. Typically one could measure the spectra of 1–3 galaxies at one time. As we have noted above cosmologists are greedy for redshifts. They want to measure as many as possible. Wouldn't it be great to measure hundreds of redshifts at one time? This has become possible through the design of fiber spectrographs. These ingenious devices still employ a long slit but use up to several hundred optical fibers to take the light from many galaxies and feed it into the slit. It is a trick to align all the galaxy images in a nice line along the spectrograph slit. Two major nearby galaxy surveys made use of these fiber spectrographs. The two degree field redshift survey was carried out using the Anglo Australian

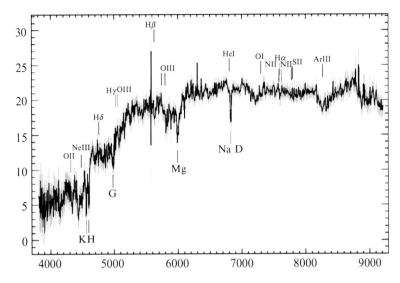

FIG. 5.1 A spectrum of an elliptical galaxy, showing how much light the galaxy gives off at different colors. The horizontal axis labels shows the wavelength of the light with *blue* on the *left* and *red light* on the *right*. The vertical axis gives the intensity of the light. The spectrum is shifted towards the *red* relative to what it would be on earth. By identifying emission features labeled in *blue*, and absorption features labeled in *red* with the known wavelengths measured in the laboratory we can calculate the redshift of this galaxy. In this case the redshift is about 0.16 (Credit: Sloan Digital Sky Survey (SDSS) Collaboration, www.sdss.org)

FIG. 5.2 The plugboard for the Sloan Digital Sky Survey fiber spectrograph. A thin plate of metal is mounted on the back of the telescope. This plate has a set of about 500 holes drilled into it at exactly the locations of 500 galaxies in the direction that the telescope is pointing. Each hole is plugged with an optical fiber, which carries the light from one galaxy down to a big spectrograph. In this way 500 galaxies can be observed at once (Credit: Fermilab Visual Media Services)

Telescope located in New South Wales, Australia. The Sloan Digital Sky Survey was carried out at Apache Point Observatory in southern New Mexico. The Sloan Digital Sky Survey project uses a thin metal plate mounted on the back of the telescope (Fig. 5.2) to hold the fibers in position. The plate has 500 holes drilled into it in exactly the locations of 500 galaxies in that area of sky. Each hole is plugged with an optical fiber which carries the light from one galaxy down to a big spectrograph. The spectrograph splits the light from each fiber into a spectrum and measures all the spectra at once.

The first step is to drill holes in metal plates. One then inserts a fiber in each hole and positions the metal plate where the image of the sky is formed in the telescope. This is hard work. You have to plug in all the fibers, note their numbers and note which galaxy's light is going down which fiber. Once you've set the whole thing up some clouds may drift by. What to do? The plates are set up for observing at a given time on a given night. Once that window is gone you have to move to the next plate.

The two degree field project used a spectrograph that carried the fibers on arms. Writing a computer program to locate 100 or more fibers on galaxy positions without the arms colliding is a tricky undertaking.

Once the data have been obtained you have to use the spectra to obtain redshifts. A CCD image may contain a million numbers corresponding to the spectra of 100 galaxies. One thus reduces a million numbers to 100 numbers, the redshifts. Most of this work can now be automated and carried out by sophisticated computer programs that produce a list of redshifts together with measurement errors and confidence levels. Measurement errors are always important in scientific experiments as we have outlined previously. Kepler argued in favor of an elliptical orbit for Mars despite the fact that a circular orbit gave a close fit because he understood the measurement errors in Tycho's data. The confidence level is a similar number. It says how certain you are that you have the correct redshift.

The galaxy surveys are thus used to produce catalogs of redshifts that can be analyzed by cosmologists to test their theories of the formation of structure in the universe. The theories specify how many clumps of a given mass should form at a specified time, thus predicting what the universe should look like today and in the past. Comparing theory to observation is an art in itself. The theorists are good at using computers to calculate in detail how the clumping of dark matter particles evolves with time. The tricky part is identifying which clumps or halos host which kinds of galaxies.

Strategies for Surveying the Universe

Before one goes to the telescope one needs a strategy or plan. You might argue that to have a plan one should anticipate what one will find. If one already knows what is to be discovered why bother, its not research. This is the dilemma of research. Research proposals are usually crafted in such a way as to argue that great progress has been made in an area of research and we can wrap it up by making this new crucial observation. However as the astrophysicist John Bahcall pointed out in making the case for the Hubble Space Telescope;

> We often frame our understanding of what the space telescope will do in terms of what we expect to find and actually it would be terribly anticlimactic if in fact we find what we expect to find... The most important discoveries will provide answers to questions that we do not yet know how to ask and will concern objects that we have not yet imagined.

> The musician Mickey Hart has put it more poetically

> Magic doesn't happen unless you set a place at the table for it.

In astronomy the key is to do something new. The pattern has been that if you don't find what you are looking for you will find something more interesting. The studies of the structure of the universe originally focused on showing that the galaxy distribution in space is not random and how the distribution deviates from randomness. The large surveys revealed that galaxies populate a large filamentary structure which we call the cosmic web, since it resembles a spider's web in some respects. The discovery of the cosmic web was completely unexpected just like the discovery of dark energy.

The Sloan Digital Sky Survey uses an unusual camera to make a photographic survey of the sky (Fig. 5.3). The camera contains six columns of CCD chips. Each column has a series of five chips, each

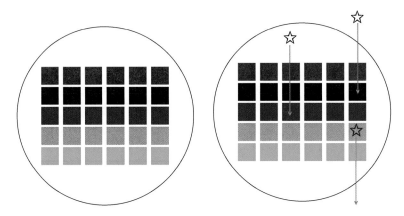

Fig. 5.3 The Sloan Digital Sky Survey CCD camera. The image on the *left* shows the six columns of CCD chips each with the five filters arranged in rows. The *right hand* image shows how a star image moves down a column during an exposure and is successively observed through different filters (Credit: Michael Carr and the Sloan Digital Sky Survey (SDSS) Collaboration, www.sdss.org)

with a different filter. The telescope is scanned across the sky so that stars move across the camera in straight lines, exactly along the columns of the CCDs. The traditional way of using CCDs on telescopes is to point the telescope at a target object. One then keeps the telescope tracking the sky so that the object remains at a fixed position on the detector. One exposes the CCD then closes the shutter then reads out the image to be stored in digital form on disk. The Sloan Survey camera operates in drift scan mode. That is to say the objects in the sky drift across the camera and the CCDs are read out in sync with the drifting. The technique produces a sharp image. An object takes about 1 min to drift across one of the CCDs. It then drifts to the next CCD in the column and so on. The result is a set of images of each object through five different color filters ranging from blue to near infrared. We can combine these five images to visualize the color of the object we are imaging.

Ideally one would want to measure the redshifts of all the objects in a galaxy catalog. In practice, it takes much longer to measure the redshifts of faint galaxies than it does to obtain an image. One thus chooses a brightness limit and an area of the sky. One then measures the redshifts of all the galaxies brighter than some limit in this specified area of the sky. Almost all redshift surveys are carried out in this manner. This method has the great advantage that it is simple. Simplicity of strategy is a virtue in cosmology. If you select galaxies for redshift measurements simply on a whim, because they look cool or whatever, it will be impossible to draw significant conclusions. The sober and systematic approach to observing is the one that pays off in the long term. It also ensures reproducibility. The authors explain what they did and how they did it and you can reproduce their results.

Redshift and imaging surveys can be divided into two main categories. The first are all-sky or wide angle surveys. Astronomers often think of the night sky as a sphere. We can describe a survey as covering some fraction of the sphere. Redshift surveys that cover large areas of sky cannot go very deep. By this we mean that we cannot observe very faint galaxies. This is because (a) there are many more faint galaxies than bright galaxies and (b) it takes more time per galaxy to obtain redshifts for faint galaxies. We can also look to much larger distances by observing faint galaxies in a small part of the sky, say half a moon diameter in size. The Keck Observatory DEEP (Deep Extragalactic Evolutionary Probe) is a case in point. This survey used a 10 m telescope to survey distant faint galaxies that had been imaged by the Hubble Space Telescope.

The large area redshift surveys tell us about the galaxy distribution on large scales. The deep redshift surveys tell us about galaxy evolution. Let us recall that as we look back over large distances we are looking back in time. The deep redshift surveys thus enable us to compare the colors and shapes of galaxies as they were a long time ago with galaxies that we see around us today. The wide angle or shallow redshift surveys tell us about the galaxy distribution nearby. As we shall see both kinds of survey provide fascinating information. Table 5.1 lists a few examples of surveys. Some of these surveys make their data available online. Over one million distinct users have accessed the Sloan Digital Sky Survey data!

Results from Surveys of Nearby Galaxies

One of the first redshift surveys designed to map the universe was carried out by astronomers at the Smithsonian Astrophysical Observatory using a 1.5 m telescope in Arizona. The survey consisted of 2,500 galaxies brighter than a certain blue color (magnitude B=14.5). Mathematical analyses showed that the galaxies were clustered.

TABLE 5.1 Examples of galaxy surveys

Acronym	Name	Website
SDSS	Sloan Digital Sky Survey	www.sdss.org/
2dFRS	2 degree Field Galaxy Redshift Survey	www.mso.anu.edu.au/2dFGRS/
2MASS	2 Micron All Sky Survey	pegasus.phast.umass.edu/
FIRST	Faint Images of the Radio Sky at Twenty cm	sundog.stsci.edu/
DEEP2	Deep Extragalactic Evolutionary Probe	deep.berkeley.edu/
VVDS	The VIRMOS-VLT Deep Survey	cesam.oamp.fr/vvdsproject/
BOSS	The Baryon Oscillation Spectroscopic Survey	cosmology.lbl.gov/BOSS/
DES	The Dark Energy Survey	darkenergysurvey.org/
LSST	The Large-Aperture Synoptic Survey Telescope	www.lsst.org/lsst/

A much more striking result was obtained by a now famous survey (known as CfA2) that was published in 1986 (Fig. 5.4). The strategy for this survey was to select galaxies in a strip on the sky. When one adds redshifts to the coordinates on the sky the volume of space surveyed looks like a slice of pizza. When the redshifts for this survey are plotted the results are striking. In such a plot one marks each galaxy as a point. One sees a clear network of filaments and voids. That is to say there are regions of space that have a spherical shape that contain no galaxies. In between these voids there are filaments and sheets. There are two other features that dominate this survey. One is the Coma cluster of galaxies, the large clump located in the center of Fig. 5.4. This is the most massive nearby cluster known. The second feature is known as the great wall. It is a filament of galaxies (a sheet or wall in three dimensions) that spans the whole survey and goes through the Coma cluster.

The fact that the CfA2 survey found this huge feature, the great wall raised an interesting question. We base our models of the universe on the assumption that the universe is homogeneous and isotropic. The CfA2 survey found a feature that was comparable in scale to the size of the survey. The fact that the slice survey was dominated by the great wall and the Coma cluster suggested the need to probe to even larger scales to get a fair sample of the

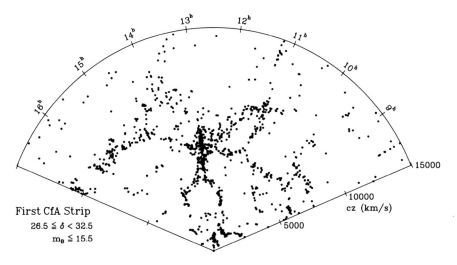

FIG. 5.4 The "slice of the Universe" that represents the first set of obser-vations done for the CfA Redshift Survey in 1985. These are spectroscopic observations of about 1,100 galaxies in a strip on the sky 6° wide and about 130° long. We are at the apex of the wedge. The radial coordinate is redshift, measured in kilometers per second. The outer arc of the plot is at a distance of about 700 million light years (Credit: Smithsonian Astrophysical Observatory)

universe. This has been a motivation for carrying out even larger and deeper surveys such as the Sloan and two degree field surveys.

The question of a fair sample can again be illustrated using a polling analogy. If I want to predict the outcome of a presidential election I could ask my neighbor who he will vote for. It would be irresponsible on the basis of one opinion to publish a statement saying that based on my research I expect Candidate A to win. I would do better if I asked a few people on my street. It would be even better to sample various neighborhoods in town and then go to neighboring towns and even other states. If I decided to poll people as a function of distance from my house, starting close by and working my way out, I could decide to stop once the results are no longer changing. That is to say I could stop when I have obtained a fair sample of the population. This is the issue facing the surveyors. The results of the Sloan Digital Sky Survey and two degree field survey show that a fair sample of the universe has finally been reached.

Both the Sloan Digital Sky Survey and the Two Degree Field Galaxy Redshift Survey produced spectacular maps of the galaxy distribution in redshift space (Fig. 5.5). The cosmic web of galaxies traced out by these surveys has been accounted for by the galaxy simulations. Notice the phrase 'accounted for' rather than predicted which is an important distinction in science.

Many scientific findings resulted from these surveys. As well as mapping out the cosmic web as traced by galaxies, the nearby galaxy surveys helped put constraints on the dark energy density, the matter density and the baryon density in the universe. One can accurately measure the star formation rate in the local universe and use this to show that the star formation rate was higher a few billion years ago than it is now. One can also study the effect of environment on galaxy properties. The colors of galaxies and the rate at which they form stars depend on the environment in which a galaxy finds itself. The surveys discovered quasars at high redshifts and made it possible to study the intergalactic gas properties at early times.

The Distribution of Mass and Light

Comparing theory with observations is somewhat like comparing two maps. One map has been given to us (by theorists Fig. 5.6) and contains the location of buried treasures, the other map has been made by us from direct observations of our surroundings (Fig. 5.5). To find the treasure we have to find features that are present in both maps and match them up. The theorists give us a a map of what the surroundings of an average observer should look like. We cannot actually find our Coma galaxy cluster of galaxies in the theorists map, but there should be somewhere a cluster of galaxies that resembles Coma. There will also be a number of halos in the simulation comparable to the number of Milky Way halos that we see in the data, but the locations will not match. Nevertheless by doing statistical studies of the two maps we can learn alot about the distribution of stars and dark matter in our universe.

The surveys provide us with a census of galaxies. We have redshifts and colors of half a million galaxies. We can use these data to estimate the mass in stars within each galaxy. The theories do not readily predict this number but they do predict how many dark matter halos of any given mass are present at any time. The

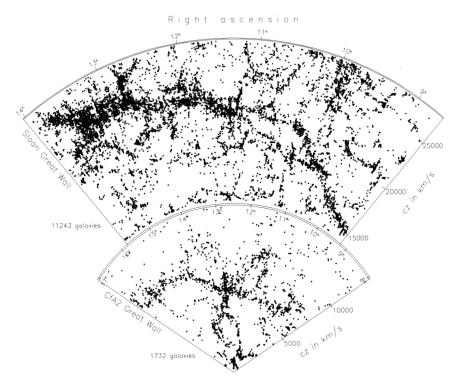

FIG. 5.5 The galaxy distribution obtained from redshift surveys. The smaller slice at the bottom shows the CfA2 survey, with the Coma cluster at the center. The upper 'slice' is section of the Sloan Digital Sky Survey in which a "Great Wall" has been identified. This is one of the largest observed structures in the Universe, containing over 10,000 galaxies and stretching over more than 1.37 billion light years (Credit: J. Richard Gott and Mario Juric, 2005, Astrophysical Journal, 624, 436, Reproduced by permission of the American Astronomical Society)

idea is to find out the mass of the dark matter halo associated with the galaxies that we observe in the surveys.

We can also obtain significant results from the observations alone. We can calculate the fraction of atoms located inside stars by estimating the total mass of stars in the survey volume. Remarkably only 3.5% of the baryons present in the universe are actually found inside stars. In other words, the processes that formed galaxies are not efficient, most of the atoms in the universe are not located in stars. When you look up at the night sky, the stars that you see account for 3.5% of the atoms in the universe which in turn account for 4.6% of the total density of the universe!

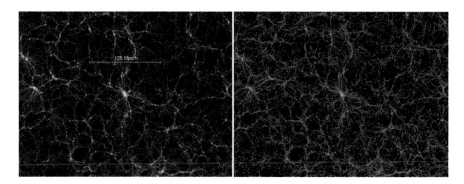

FIG. 5.6 The model universe from computer simulations. The *left-hand* image shows the dark matter distribution in the cosmic web. The *right-hand* image shows the expected distribution of visible light when the halos in the *left hand* image are populated with stars following a plausible star formation recipe (Credit: Volker Springel (Heidelberg University) and the Virgo Consortium)

Most stars in the universe are found in galaxies with a similar stellar mass to the Milky Way. There is a problem with trying to match up observed galaxies in a survey with halos in a simulation. The model produces too many low mass and high mass halos if we fix it so that the right number of Milky Way mass galaxies are produced.

If we remove the assumption that the amount of visible matter (stellar mass) is proportional to the amount of dark matter in a galaxy we can match the numbers of observed galaxies in the maps to the observed number of halos of given mass in the simulation. We can then estimate the relative amounts of visible and dark matter in galaxies of different mass.

Review

In this chapter we saw how systematic surveys of the sky have revealed the largest known structures in the universe. From the field of biology a book entitled "The Beak of the Finch" illustrates the value of survey work in science. It describes the work of scientists on a small island in the Galapagos. They studied over a period of decades the population of finches on the island and got to see evolution in action. They showed how a difference in beak size of a fraction of a millimeter could mean life or death for a finch during hard times when food is scarce. This study was able to see natural selection taking place in real time in a bird population. The work consisted of repeating exactly the same observations year after year, measuring beak sizes and putting metal bands on birds. However the insights obtained from this work are intriguing and wonderful.

The same is true in astronomy. Many new insights come as a result of survey work. The work is inherently boring and repetitive. We point our telescope and measure some galaxy redshifts then point it somewhere else and measure some more redshifts, on and on, night after night. It is out of such work that the realization came that there are enormous voids in space, regions empty or almost empty of galaxies. Surveys are important because we study the universe in a controlled matter. It is hard to draw conclusions about a sample of galaxies that were selected because they seemed interesting to the author, on the other hand if we select all galaxies in the sky brighter than some limit and study their properties our conclusions will have some meaning.

As with so much of astronomy, this field has made great use of technological developments. With new spectrographs we can measure the redshifts of 500 galaxies in an hour or so. It seems every decade or two things improve such that what used to take a few months now can be done in one night of observing. This is of importance because we need large numbers of galaxy redshifts to map out the matter distribution in the universe.

There is a simple satisfaction at simply knowing our surroundings in space. When I was a small child, the universe was the neighborhood on the scale of a few hundred yards. I remember an old house falling apart in a wooded area nearby that was forbidden territory. That was my frontier beyond which lay the unknown. Eventually I expanded my horizons well beyond this. In that same way I find it interesting to simply know where we live on a cosmic scale. Over there in Coma is a huge cluster of galaxies, over in Perseus lies a big filament of galaxies that covers half the sky and so on. By making these maps we can study our neighborhood and compare it to others.

We have seen that we can map out the universe nearby over a large area of the sky, and we can probe the distant universe by studying distant galaxies in a small part of the sky. Both these approaches are equally important. The nearby surveys map out the mass nearby. The distant galaxy surveys give us the chance to study the evolution of galaxies. We can see if the galaxies looked different in the past. What was the rate of star formation at high redshifts compared to today in a given volume of space? Do galaxy shapes and sizes look different at redshift one than they do today?

The amount of available scientific data doubles every year. In astronomy this is due more to the increase in the available detector area than the actual collecting area of telescopes. Part of the challenge is to make the survey data available to the astronomical community and also to the general public. The Sloan Digital Sky Survey developed a web tool called Skyserver to access the data. This tool has had about one billion hits in the last decade. It has had over one million distinct users. An associated project was the galaxy zoo project where 40 million galaxy classifications were made by the general public. Three hundred thousand people participated in this project. The data explosion resulting from astronomical survey projects requires new tools to extract the information. This is one of the new frontiers of research in what has been called the age of surveys.

The fact that galaxies are distributed in a cosmic web extending for billions of light years raises the question of origins. How did the cosmic web come into existence and why do some huge regions of space contain no galaxies at all? It is these questions that we turn to in the next chapter.

Further Reading

A Grand and Bold Thing: An Extraordinary New Map of the Universe Ushering In A New Era of Discovery. Ann K. Finkbeiner, New York, Simon & Schuster. 2010.

Mapping the Universe: The Interactive History of Astronomy. Paul Murdin, London, Carlton Books, 2011.

6. How Did Galaxies Come into Existence?

If the matter was evenly disposed throughout an infinite space, it could never convene into one mass; but some of it would convene into one mass and some into an other, so as to make an infinite number of great masses scattered at great distances from one to another throughout all that infinite space. And thus might the Sun and fixed stars be formed.

Isaac Newton, Letter to Richard Bentley, 1692

How can we understand the origin of galaxies? The problem is not unlike that of the origin of life on earth. We see a wide variety of species in the world today. The fossil records tells us that species were different in the past. At very early times only very simple single celled organisms existed. Biologists studying the origin of life must ask themselves how so much complexity and diversity emerged from simple beginnings. They must also ask what got the process going in the first place.

The largest structure revealed to date is a huge filament consisting of thousands of galaxies. This structure known as the Sloan Great Wall is about 1.4 billion light years across. Like scientists studying the origin of life on earth, astronomers want to understand the origin of galaxies and the large structures that they are part of. We want to understand how so much complexity emerged from the early universe which was almost featureless.

Observations of the cosmic background radiation tell us that when the universe was a few hundred thousand years old no stars or galaxies existed, the density of matter was almost perfectly smooth. The density of matter varied from one place to the next by less than one part in a 100,000. How then did the complex structures and galaxies described in the previous chapter emerge from the early universe?

G. Rhee, *Cosmic Dawn: The Search for the First Stars and Galaxies*, Astronomers' Universe, DOI 10.1007/978-1-4614-7813-3_6, © Springer Science+Business Media, LLC 2013

We present the solution to this puzzle in this chapter. We begin by discussing the mechanism by which gravity amplifies small density variations present in the early universe. To properly model the emergence of the cosmic web we need to use computers. We start with an almost totally smooth matter distribution and use computers to model how this distribution changes due to the effect of gravity in an expanding universe. We discuss how by adding gas to the dark matter calculations we can gain more insight into the process of galaxy formation.

The Gravitational Instability: How the Rich Get Richer

What would happen to a part of the universe that was very slightly denser than its surroundings when the universe was only 400,000 years old? This region will expand slightly slower than a region of average density because of the gravitational force caused by the small excess of matter. Since the region expands more slowly, the density in this region is dropping slower than that of its surroundings. This in turn means that the density of the region relative to its surroundings is increasing.

As an analogy let us consider two cars. Car A uses gas at a rate of 10 miles per gallon. Car B uses gas at a rate of 20 miles per gallon. The two cars fill up their gas tanks with 10 gallons of gas and start on a journey. After 20 miles, car A has 8 gallons left and Car B has 9 gallons left. Car B has about 1.1 times as much gas left as car A. Twenty miles further car A has 6 gallons left and car B has 8 gallons left, car B now has 1.3 times as much gas in its tank as car A. By the time the two cars have gone 8 miles car A has three times as much gas in its tank as car B. Although both cars have less and less gas in their tanks, the relative amount of gas of car B to car A keeps increasing.

Our overdense region keeps getting less and less dense while its density relative to its surroundings keeps increasing. In the universe as in capitalism, the rich eventually get richer and the poor get poorer. Gravity thus acts as an amplifier, it enhances small differences in density that were present in the early universe. Regions that are initially slightly denser than their surroundings will subsequently form galaxies and clusters of galaxies. Conversely, regions of space that are initially less dense than average will become more and more empty forming the voids that we saw in the galaxy maps shown in Figs. 5.4 and 5.5.

At some point an overdense region will stop expanding and collapse to form a bound object such as our galaxy. There are many regions of varying sizes going through this process. The regions which gave rise to the first galaxies collapsed long ago at redshifts of 12 or higher (less than 400 million years after the Big Bang). Some of the larger regions bigger than clusters of galaxies are only just collapsing today.

To summarize, gravity causes slight variations in the density of the universe to increase with time. We believe it is this effect that leads to the formation of galaxies from a universe that is almost featureless. Of course the story does not end there. We would like to explain the properties of the cosmic web of galaxies. This simple model of the evolution of isolated overdense regions does not explain how a feature such as the Sloan Great Wall which is itself made up of thousands of galaxies could come into existence. To answer that question we use computers to simulate the gravitational interaction of hundreds of millions of particles.

Simulating the Universe Inside a Computer

Astronomers follow the evolution of dark matter using computer simulations. The calculations are based on the assumption that gravity is the only way that the particles influence each other. We can calculate the effects of gravity and predict the trajectories of particles in space using computers. The simplest calculation involves just two particles. At any given time if we know the positions of the particles we can calculate the strength of the gravitational force between them. If we also know the speeds at which the particles are moving, we can predict where the particles will be at any time in the future. If no force acts, the solution is very simple. The particles just keep going forever with whatever velocities they had initially. If the particles have mass they will affect each other's motion due to the force of gravity. By calculating the force of gravity we can calculate the acceleration of the particles, that is to say, how their motion changes. By jumping from the present to a short time in the future we can predict the new velocities of the particles using Newton's laws of motion. Knowing the velocities we calculate the new positions of the particles. So the simulation process consists of a repeating set of instructions. Note the location and velocities of all the particles, compute the forces, calculate the new velocities and location of the particles and repeat the process.

We can apply this method to the Earth's orbit around the Sun. We give the program starting positions of the Earth and the Sun and the direction of the Earth's motion. We can then compute where the earth will be a year and a half from today. The program cannot compute this in one step so it breaks down the task into a series of steps. We could, for example, compute the forces, velocities, and particle positions 500 times per orbit. If we did it 5,000 times per orbit the answers would be even more accurate.

How did Newton figure out the orbits of the planets since he had no computer? The analysis of motion in the solar system is considerably simplified by the fact that by ignoring the effect of the other planets on the Earth we get a pretty accurate answer. The reason Sir Isaac came up with the answer, is that this so-called two body problem can be solved exactly using equations. We can circumvent the computer because we can write the equation that describes an ellipse as a solution to the equations of motion of the planets. To put it simply, for certain problems we can write down the answer as a formula, for others we cannot. If I give a physicist the positions and velocities of 1,000 particles she cannot write down a simple formula for their future trajectories. We have to use a computer to perform the calculation. We are thus forced to break this problem down into a series of steps as described above.

With our computer code we are not limited to problems that have simple solutions. We can turn gravity into a great video game. Just specify the starting conditions, also known as initial conditions, and let it roll. We could have three particles of equal mass for example. The program then calculates the force on particle one due to particle two and similarly the force on particle one due to particle three. With appropriate permutations we do the same for particles two and three. Notice that the number of force computations has now grown. In fact, the number of computations increases as the square of the number of particles in the simulation. Thus if we double the number of particles we have to spend four times as much time on the computations.

Astrophysicists need to perform simulations with very large numbers of particles. The reason for this is that the structure of galaxies is influenced by processes occurring on much larger scales. We want to see how the Sloan Great Wall was formed while at the same time seeing how the galaxies that make up the wall were formed. In order to get the full picture we need to know what happens on a scale of a billion light years as well as hundreds of light years at the same time. With the computing power of computers we are able to do this. An example of a state of the art supercomputer is the Blue Waters supercomputer funded by the National Science Foundation. When completed it will be able to carry out 10^{15} calculations per second. The iphone 4 can carry out about 40 million (4×10^7) calculations per second. The state of the art for dark matter simulations is currently the Bolshoi simulation which calculates the motion of about 10 billion particles of dark matter in a box 1 billion light years on a side.

If the number of particles is 10 billion, 10^{10}, then to compute the force on one of the particles we must carry out 10 billion minus one calculations. To compute the forces on all the particles we must carry out 10^{20} computations. One of the objectives of people writing the computer programs that carry out these simulations is to see if they can speed up the computations. One ingenious trick is to note that particles far away from the point at which we are trying to calculate the gravitational force can be treated as one particle if they are clumped together. For example I can treat the planet Pluto and its moon as one object when I calculate their pull on me. Pluto and its moon are so close together compared to their distance from me that they pull me in essentially the same direction so I can treat them as one point object. One might argue that this is an approximation and that we want the exact answer in science. The art of being a good astronomer is the art of approximation, of knowing what simplifying assumptions to make to render a problem more tractable. We do this in everyday life. When someone asks me how long a drive I have to work, I say about 20 miles. I could say 22.145 miles but I would get some strange looks. I say 20 or so miles because that is the accuracy that is required in ordinary conversation..

We know what the universe looks like today from the maps shown in the previous chapter. We know how dark matter clumps will grow thanks to the simulation techniques described above. We can only run our simulations forward in time so we have to know the dark matter distribution at early times to set up the starting point of our simulations.

How to Start a Model Universe

To start a computer simulation we must specify the initial velocities and positions of the particles whose evolution we are to follow from the past to the present day. The initial conditions can be thought of as specifying how the density of the early universe varied from one location to the next. We characterize the variations in density using a statistical description and then make a specific density distribution from the description. The idea behind this is that we can run several versions of a model universe that are statistically similar to see the effect of random variations in density. These models are the same on average but differ in the exact locations that the galaxies form.

To start our simulation we need to know the size of the small variations in density in the universe before galaxies and stars formed. We have a reliable estimate from observations of the cosmic background radiation intensity from one part of the sky to the next. The WMAP satellite has accurately measured these. From these variations we can estimate the variations in density of the dark matter from one location to the next when the universe was only 400,000 years old. In other words we have a reliable way to determine the starting conditions for our simulation when the density variations were very small. We want to know if we can reproduce all this complex structure we see today using this set of very simple initial conditions. Amazingly the answer seems to be yes, which means we understand something about the role of gravity in structure formation. Note however that we do not understand the origin of the variations. We do have some plausible ideas as to their cause but we do not have any direct evidence for the mechanism.

Once we have run a dark matter simulation we need to see if it matches the real world. Unfortunately we only have observations of the light emitted by galaxies, not the dark matter itself. We thus have to first find a way to estimate where the stars and galaxies will form in the simulation and then compare these light-emitting regions to the maps produced by the galaxy surveys.

Matching the Models to the Real World

The simplest way to compare the distribution of galaxies in the real world and the simulations would be to compare pictures and judge by eye. This is not as naive as it sounds, the eye is actually very good at discriminating between pictures. I don't think we yet have a computer algorithm that can tell the difference between a genuine Rembrandt and a fake. A more objective method is to calculate some numbers from the data and see if they match numbers computed from the models. The galaxy distributions form a sponge like structure known as the cosmic web. It is not easy to quantify sponginess to compare two sponges but it can be done.

For example we can calculate how many galaxy clusters are present in a galaxy survey and compare that with the number of clusters seen in a simulated volume of the universe. We can use the models to mimic observational data. We then compare the results by applying statistical tests to real and model data and see if they match.

At the other extreme we know that there are large regions of space (tens of millions of light years across) that are empty of galaxies. We can see if we find similar sized voids in the simulations.

In the universe, the rich get richer and the poor get poorer. By this we mean that a structure that is only very slightly denser than its surroundings in the early universe will become increasingly overdense as time passes. Similarly, a region that is slightly less dense will become increasingly less dense. Gravity like capitalism accentuates differences. Another feature of the gravitational instability is that the overdense regions which are not perfectly spherical become increasingly elliptical or elongated with time. This explains the formation of filaments that constitute the cosmic web. The underdense regions on the other hand become increasingly spherical with time. The voids or empty regions observed in galaxy surveys do in fact turn out to have a roughly spherical shape. The matter distribution resembles a sponge. The low density regions, the holes in the sponge, are somewhat spherical while the high density regions form a connecting network of filaments and sheets with large massive objects forming at the intersections of filaments.

Our ability to make simulations has improved at a similar pace to our ability to carry out observations. Computers have enabled us to carry out surveys of galaxies over large areas of the sky and to great distances in reasonable amounts of time. Similarly, computers have enabled us to carry out simulations with sufficient resolution that we can say something about galaxies and their distribution in space. Astronomy is driven by improvements in technology as well as new physical insights. The improvements in technology show no signs of abating. Computers keep improving at a staggering pace and new telescopes and better detectors enable us to make observations of better quality and quantity.

Crunching Big Numbers: The State of the Art

The Virgo consortium, an international collaboration of a dozen scientists in the UK, Germany, Netherlands, Canada the USA and Japan has produced some of the most detailed computer simulations. To carry out these calculations they use computers that have more than 3,000 cores with 14 TB of memory and 600 TB disk storage. The simulations can take anything from a week to several months on a 100 or more processors.

We can run a large-scale simulation and zoom in on the result to find interesting examples of clusters of galaxies (Fig. 6.1). To get a better picture we can rerun the simulation of a smaller volume centered on the interesting cluster. There are a number of properties of galaxies in clusters that differ from galaxies outside clusters. For example the proportion of elliptical galaxies relative to spiral galaxies is much higher in clusters. The star formation rate of galaxies in clusters seems to be different at different redshifts and so on. Almost all the galaxies in clusters are less than a tenth of a cluster diameter in size. We want to have the big picture (cluster) but also see the small picture (galaxies) in sufficient detail to see if it is influenced by the big picture. In order to do this we need more particles in the cluster region than are

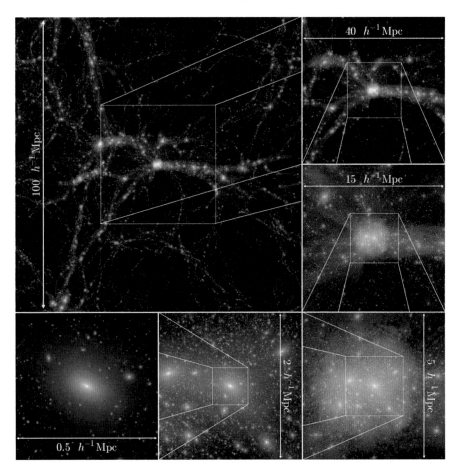

FIG. 6.1 A zoom through the Virgo consortium Millennium-II Simulation of dark matter evolution. The images all correspond to a moment in time. The dark matter density is color coded so that high density regions are *white* and low density regions are *black*. The simulation follows about 10 billion particles in a cube roughly 500 million light years on a side. The units shown in the figure are Megaparsecs where 1 parsec is about 3 light years. The *bottom* three boxes zoom in on a structure comparable in mass to the Coma cluster an aggregation of many thousands of galaxies. The images show how halos are embedded in larger structures that constitute the cosmic web (Credit: Michael Boylan-Kolchin, Volker Springel, Simon D. M. White, Adrian Jenkins, and Gerard Lemson (2009) Monthly Notices of the Royal Astronomical Society, 398 1150B, by permission of Oxford University Press on behalf of the Royal Astronomical Society)

provided by the large scale simulation. We thus run a smaller sized simulation with as many particles as the big simulation. The large scale simulations consist of a cube 700 million light years on a side, the smaller simulations correspond to a cube 2 million light years on a side.

If you search the world wide web under the heading "N-body simulations" you will find many sites that contain images and movies produced using these computations. The movies give insight into the growth of the three dimensional structures as they form. Looking at this material you see structure arise out of the almost uniform early universe. It is remarkable that astrophysicists have solved key elements of this problem. The more mathematically minded and adventurous readers can find computer codes on the web and run their own simulations. For example, you can make an elliptical galaxy and have it collide with another one and watch the results. The point is that the codes have been written and all you have to do is to type in a few simple instructions to run the program and get results.

So far in this chapter we have emphasized the role of gravity in determining the large scale structure of the universe. The visible parts of the universe are visible because of the presence of gas that emits and absorbs light. To understand galaxy formation we must include the physics of gas. This involves more than gravity, since the gas, unlike the dark matter, responds to pressure and can dissipate energy by radiation.

The Effect of Gas: A Study in Dissipation

None of the simulations described above can produce stars and galaxies. The models simulate the collapse of lumps of dark matter due to gravity. They can also include the behavior gas, on large scales. In order to compare simulations with observations we need a scheme for labeling as visible some of the gas in the simulations. One has to specify under which conditions (density, temperature, composition) gas will turn into stars. Gas made of atoms and molecules is different from dark matter in that it can emit and absorb light. The details of these physical processes are quite complex and not always well established. One uses a prescription rather than actually simulating the star formation

FIG. 6.2 Which galaxy is real? A galaxy in a simulation (*left*) appears in all respects identical to a real galaxy (*right*) and background image from the Sloan Digital Sky Survey Collaboration. The simulation includes a star formation recipe and calculates the effect of interactions between the stars and the gas (Credit: Reprinted by permission from Macmillan Publishers Ltd: Nature, Geha, M. 2010, 463, 167, copyright (2010))

process. This is currently the only way to compare cosmological simulations to observations. This simplistic approach is supported by observations of galaxies which reveal a threshold gas density for star formation. The star formation rate increases as the gas density increases. These relations between star formation and gas density are found for very nearby galaxies; we do not know if they hold at high redshifts. Simulations including gas are made for cubes 300 million light years on a side with about 300 million dark matter particles and 300 million gas particles initially. As the simulation is run, the dark matter particles remain unchanged but the gas particles get converted to stars. The smallest structures that can be seen in these simulations are about 20,000 light years in size. We are still a long way from simulating the formation of galaxies and the formation of stars inside those galaxies using the basic equations of physics, hence the need for star formation recipes. Current simulations can produce quite realistic looking galaxies. Figure 6.2 shows a simulated galaxy including gas physics and star formation and a real galaxy image.

A second approach is to run the dark matter simulation first and then retroactively add the effects of gas. Dark matter halos form by the merging of smaller halos to form larger ones. For a given halo that exists at the end of a simulation we can construct a merger history, namely locate all of the halos in the simulation that contributed mass to the halo in question. One then adds gas to these halos in the form of hot gas, cold gas and stars. Simple prescriptions are then used to determine how the baryons flow between these three components. For example, stars form out of the gas while supernova explosions return stellar baryons to the gas phase. By keeping track of all these processes one can predict what the stellar population of the final halo will be. It is then possible to predict the observables such as the colors and luminosities of the galaxies.

These models produce catalogs of galaxies whose properties can be compared to the properties of galaxies in the surveys. For example we can measure the relative numbers of bright and faint galaxies in the models and the surveys. There are many more faint galaxies than bright galaxies and at the bright end there is a cutoff beyond which there are no bright galaxies. The relative numbers of bright and faint galaxies can currently be estimated out to redshifts of about eight (600 million years after the Big Bang) using Hubble Space Telescope observations.

Cosmological simulations usually explain rather than predict astronomical observations. It is as if I tell you that I under-stand the stock market. To prove it, I can claim to explain all the available data on the S&P 500 index. However, I still have no idea what the index will do tomorrow. Physical science should be able to do better than that. Nevertheless the simu-lations are very useful as a guide to the complexities of galaxy formation.

Prediction is not easy. One does not need to discuss simu-lations to make this point. The X-ray emission from clusters of galaxies could easily have been predicted before its discovery. The physics is so simple that it is set as a question in an introductory astronomy text. Yet, no one did predict the existence of a hot intracluster medium. Predicting the future is more difficult than predicting the past.

The simulations have produced some beautiful results. They enable us to visualize the process of dark matter halo formation as it takes place. In cosmology, observations and theory are intertwined to their mutual benefit.

Why Do Galaxies Rotate?

The rising and setting of the Sun, the phases of the Moon and the seasons recur regularly because of the rotation of our planet and its moon. The moon rotates around the earth, the Earth revolves around the Sun, and of course the Earth revolves on its axis. As we have seen, the whole solar system revolves around the center of our galaxy. How did things start rotating in the first place?

Scientist like to reason by analogy. If you have looked at a river flowing, you may have noticed the presence of eddies and turbulence in the flow of the water. Some researchers have argued that turbulence in the early universe would result in rotation. However, when examined in detail, this picture dues not work. The prevalent view of rotation in galaxies is that it arises due to tides. Tides arise because the force of gravity weakens with distance. Thus, the gravitational pull of the Moon on the Earth is stronger on the side of the Earth closest to the Moon. This effect results in a bulge in the oceans The surface of the Earth itself is distorted by tides of about one foot. That is the surface of the Earth rises and falls by about one foot every 12 h. This effect has been measured using a network of radio telescopes and a technique called Very Large Baseline Interferometry. The tidal effect is most pronounced for Io, one of Jupiter's satellites where the tidal force due to Jupiter heats up the planet and causes the formation of active volcanos. The surface of Io rises and falls by several hundred feet due this effect.

What of tides in the early universe? Imagine an egg-shaped clump of dark matter. This clump is not isolated. There is another clump of dark matter not too far away from it. Let us consider the gravitational force on our clump due to its neighbor. The end of our egg shaped clump that is closest to the neighbor will experience a stronger gravitational pull from the neighbor than the end that is furthest. This is what we mean by a tide. This effect will set the clump spinning.

The same effect could occur if you were to fall feet first in space towards a black hole or some such dense object. If your body were aligned perfectly with a line drawn from you to the center of the black hole, your body would remain aligned as you fall. You would experience a tidal force, since your feet being closer to the black hole want to move faster than your head. What if you are not perfectly aligned? What if your body is tilted at an angle to the line from you to the center of the black hole? The effect of the tidal force will be to set you spinning. One final analogy. Imagine a canoe with a rope at each end that lies parallel to the shore of a lake. You and your friend both pull on each rope and the canoe slowly moves towards the shore. If you pull harder than your friend, the canoe will start rotating, as well as moving towards the shore.

Elliptical galaxies rotate very little while spiral galaxies rotate quite fast. Our calculations show that we can account for the spin of elliptical galaxies, but not that of spiral galaxies, using the tidal argument.

The formation of rotating disks in spiral galaxies is more complex. The galaxies grow by mergers and the structure of the disk that eventually forms depends on how the star formation affects the gas during merger events. Mergers can also destroy disks and lead to the formation of elliptical galaxies as shown in Fig. 6.3. Simulations can currently form disk galaxies that look similar to actual spiral galaxies (Fig. 6.2) but there are still many elements of the formation process that need to be understood.

The Formation of the Cosmic Web

Galaxy maps reveal galaxies located inside filaments and clusters at the meeting points of these filaments. There are also walls or sheets of galaxies located within this web. The largest of these structures span hundreds of millions of light years. A similar cosmic web forms in the simulations of the formation of galaxies. Unlike star formation, the web is not put in by hand in the simulations, it emerges naturally from the very small variations in density from one location to the next that were discovered by COBE and WMAP satellite images of the early universe.

One can think mathematically of the density variations as being waves on an ocean. To get the true shape of the surface of the ocean, we add up waves of differing wavelengths. The waves are

FIG. 6.3 The merging of two spiral galaxies like our own Milky Way. As the two galaxies approach (*upper left*), on their first close pass, they sideswipe each other, throwing out long tails of stars and gas (*upper right*). They then move apart before gravity pulls them back together again and their centers merge, when the stars at greater distances from the center are tossed into random orbits (*lower left*). Eventually the merged spiral galaxies form an elliptical galaxy (*lower right*) (Credit: Patrik Jonsson, Greg Novak and Joel Primack, University of California, Santa Cruz, 2008)

in fact specified by two numbers, their wavelengths, the distance from crest to crest, and their amplitude, the height of the waves. We can estimate the amplitude of the waves from observations of the cosmic background radiation. Unlike the two dimensional surface of the ocean, we are dealing with a three dimensional density distribution. So, we add up the waves in three dimensions. In this way we produce a model that gives us the starting value of the density of dark matter at any location in space. We can then use our computers to calculate the density at any time in the future. We run our simulation and stop it when we think we have the best match with the observations. Figure 6.4 illustrates the formation of the cosmic web in a simulation using a series of snapshots of

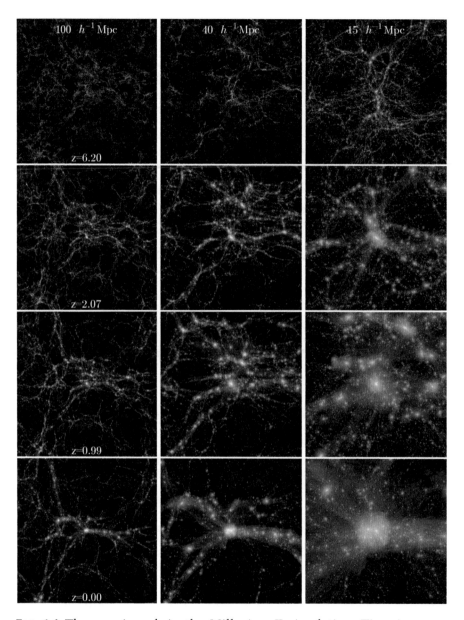

FIG. 6.4 The cosmic web in the Millenium-II simulation. Time increases from top to bottom in each of the three columns. Column 2 figures are a zoom in of column 1. The third column is a further zoom in. The z numbers label the redshifts corresponding to the time in the simulation when the snapshot was taken. The *top row* is the dark matter distribution about 1 billion years after the Big Bang. The *bottom row* is the dark matter distribution today. The image box size is expanding at the same rate as the universe (Credit: Michael Boylan-Kolchin, Volker Springel, Simon D. M. White, Adrian Jenkins, and Gerard Lemson (2009) Monthly Notices of the Royal Astronomical Society, 398 1150B, by permission of Oxford University Press on behalf of the Royal Astronomical Society)

the particles. It is remarkable that this simple procedure produces results that match the observations. Starting with observations of light from a few hundred thousand years after the big bang we can explain key features of the galaxy maps that we have made if not the details of the galaxies themselves.

Review: What the Theorists Taught Us

The calculations show that a region that is slightly denser than its surroundings will grow increasingly overdense with time until it finally collapses to form an object such as a galaxy. We can follow this process initially with simple equations that we can solve with pencil and paper. To create a more realistic picture with many dark matter halos of various masses forming and interacting we require powerful computers. The results reveal in detail the structures that form through the action of gravity. We can measure how their density varies with distance from the center of the object. We can study the dark matter halos as they form. We can determine the number and shape of these halos. We can also study how the halos are affected by collisions and mergers. Finally, we can measure distribution of these halos in space.

The results depend on the very small, density variations that we start with. We describe these density variations mathematically using waves of varying wavelength and amplitude. The density variations of large wavelength had smaller amplitude than the smaller wavelength density variations. One could use a sound recording as an analogy. The density variations on large scales correspond to the bass frequencies and the small scale variations correspond to high frequencies. The idea is that low frequencies are less prominent in the recording than the high frequencies. The role of gravity is then equivalent to turning up the volume of all the frequencies. This simple idea is valid until the variations in density are comparable with the density itself at which point this simple picture breaks down and the cosmic web and many collapsed halos start to form.

For many years, there were no data to support this picture. But, in the early 1990s, NASA's Cosmic Background Explorer Satellite followed in 2001 by NASA's WMAP satellite provided evidence of variations in the cosmic background radiation intensity from one place to the next. This is a tremendously important finding for cosmology. It reveals how large the matter density variations were at a finite time in the very distant past. This confirms that our simple view of galaxy formation is at least partly correct and enables us to put observed numbers into our theoretical speculation. In the next chapter we present the results of observations of the cosmic background radiation that are fundamental to cosmology.

Further Reading

Introduction to Cosmology. B. Ryden, San Francisco, Addison Wesley, 2003.

How Did the First Stars and Galaxies Form? A. Loeb, Princeton, Princeton University Press, 2010.

Theoretical challenges in understanding galaxy evolution J.P. Ostriker and T. Naab, Physics Today, volume 65, p43, 2012.

7. The Weight, Shape, and Fate of the Universe

Madame Dieterlen gave me coffee in her caravan. I asked her what traces the cattle herders of the Sahel would leave for an archaeologist. She thought for a moment and answered, "They scatter the ashes of their fires. No, Your archaeologist would not find those. But the women do weave little chaplets from grass stems, and hang them from the branch of their shade trees."

Bruce Chatwin, The Songlines

In order to estimate the size, luminosity and mass of galaxies we need to know the Hubble constant and the density of the universe. In recent years, these numbers have been measured to better than 1% accuracy. This is a remarkable improvement on the situation 20 years ago when the Hubble constant was not known to better than a factor of 2.

Three observations taken together have made this improvement possible. The first is the observation of the variations in brightness of the cosmic background radiation. The second is the observation of distant supernova explosions. The third is the observation of the clustering of galaxies. The attempt to better constrain the density of the universe using supernovas led to the remarkable discovery that instead of slowing down as we might, expect the expansion of the universe is in fact accelerating. This discovery implies the existence of a new force in the universe that astronomers call dark energy.

G. Rhee, *Cosmic Dawn: The Search for the First Stars and Galaxies*, Astronomers' Universe, DOI 10.1007/978-1-4614-7813-3_7, © Springer Science+Business Media, LLC 2013

Echoes of Creation: Discovery of the Cosmic Microwave Background

Remarkably most of the light in the universe was not emitted by stars and galaxies but comes from the cosmic background radiation, the afterglow of the Big Bang (see Chap. 2). This background radiation was discovered serendipitously by Arno Penzias and Robert Wilson, two researchers working at Bell Laboratories. The antenna they used had been constructed for the purpose of satellite communication. When astronomers look at data they distinguish between signal (such as light from a galaxy) and noise (random signals from other sources such as detector electronics and background light from the night sky). One man's noise can be another man's signal. If you are looking for the best radio frequencies for satellite communications, any diffuse radiation from the night sky is an annoyance and is classified as noise. If you are an astrophysicist looking for the light that has been traveling 13.7 billion years since the Big Bang, that diffuse radiation is most definitely signal.

In the early 1960s astronomers thought that there was only one significant source of background radiation, the Milky Way galaxy. Penzias and Wilson were thus setting out to measure to high accuracy the background radiation of our galaxy at radio wavelengths. They started their observations at a wavelength of 7 cm where they expected to see no signal. This would be a way for them to measure the noise in their antenna and also noise from the Earth's atmosphere. They were surprised to detect a significant amount of radiation at 7 cm and checked the antenna and the instruments to make sure no mistake had been made. In the end they had to conclude that it looked as if they were detecting background radiation emitted by matter at a temperature of $3.5°$ K. We call it background radiation because it is diffuse, it seemed to come from all over the sky. We see this effect when we look at the night sky from a big city. It is much harder to see faint stars because of all the city light that makes the sky very bright. Penzias and Wilson did not know the origin of the radiation they had found, they had in fact detected the afterglow of creation.

They found this out by talking to astronomers. Penzias called up Bernard Burke at MIT, and mentioned the issue of the background radiation he had detected. Burke had heard rumors of a talk by Princeton University astrophysicist Jim Peebles concerning the anticipated existence of background radiation left over from the Big Bang. Penzias and Wilson published their observations in a paper entitled "A measurement of Excess Antenna Temperature at 4,080 Mc/s". This work earned them the Nobel prize for physics in 1978. An accompanying paper by Dicke, Peebles, Roll and Wilkinson gave the explanation for the origin of the background radiation.

Steven Weinberg in his book *The First Three Minutes* points out that long before the year of discovery (1965) it would have been possible to both predict and detect the existence of this radiation. In fact, both the prediction of the radiation and the measurements proving its existence existed since the 1940s but no one put two and two together. The chain of argument is quite simple once one knows the helium abundance in nature. Ten percent of the atoms in the universe today are helium atoms. This number is determined by the neutron to proton ratio at the time of nucleosynthesis. This neutron to proton ratio can be used to infer the approximate temperature and density at which nucleosynthesis took place. By observing the density of baryons today we can determine by how much the universe has expanded since nucleosynthesis. The known amount of expansion enables us to determine the temperature of the radiation today. This rough argument would enable us to predict the temperature of the background radiation to lie between 1 and 10 K.

Interestingly in 1948, there was a prediction made of the existence of cosmic background radiation with a temperature of 5° K. This work was done by George Gamow, Ralph Alpher and Robert Herman. George Gamow was born in the Ukraine in 1904. Prior to settling in the US, he moved from one renowned center of European physics to the next. He worked at the University of Gottingen, the Cavendish Laboratory in Cambridge and the Institute of Theoretical Physics in Copenhagen. Gamow was named Professor of Physics at Leningrad University in 1931. He escaped Stalin's oppressive regime in 1933 and moved to the United States, working first at George Washington University and later at the University of Colorado at Boulder.

It would have been possible to detect the cosmic background radiation as early as the mid 1940s. Why did this not happen? Weinberg lists three reasons. First, the Big Bang theory of the origin of the elements ran into a number of problems as it was first formulated. It purported to explain the abundances of all the elements. But, we know today that elements such as carbon and heavier elements are made in stars and supernova explosions. It was thus clear at the time that the Big Bang theory could not explain the origin of all the elements. As Weinberg puts it;

> The Big Bang theory of nucleosynthesis, by trying to do too much, had lost the plausibility that it really deserved as a theory of helium synthesis.

The second reason the background was not detected earlier was that theorists did not realize it could be detected. The predicted temperatures for the background placed most of the emission in the radio part of the spectrum and radio astronomy was still a young science at the time. The third reason stated by Weinberg is that the Big Bang theory was not taken seriously at the time;

> This is often the way it is in physics – our mistake is not that we take our theories too seriously, but that we do not take them seriously enough. It is always hard to realize that these numbers and equations we play with at our desks have something to do with the real world ... The most important thing accomplished by the ultimate discovery of the 3 K radiation background in 1965 was to force us all to take seriously the idea that there *was* an early universe.

In fact, the cosmic background radiation had been detected in 1941, but not recognized for what it was. Adams and McKellar studied the absorption of light by cyanogen molecules in gas clouds and concluded that the molecules were interacting with light having an effective temperature of 2.3 K.

When we teach science we emphasize its successes. As much if not more can be learned from its failures. No doubt historians of science will look back on the present era and ask how astronomers could have failed to ask the questions that seem 'obvious in retrospect'.

Detecting the Cosmic Background

The cosmic background radiation is light that is emitted by matter at a uniform temperature. An oven heated to a given temperature will emit radiation. We could make a small hole in the oven and observe the radiation that is emitted. Kitchen ovens do not emit visible light but they do emit light in the form of infrared radiation. We are in some sense living in an oven at 3° above zero.

The spectrum of the cosmic background radiation varies smoothly from one wavelength to another (Fig. 7.1). The correct mathematical formula describing how the intensity changes with wavelength was found by Max Planck about 100 years ago. The detailed study of this problem led to the birth of quantum physics. The intensity peaks at a wavelength of about a tenth of a centimeter. This is because the radiation has been greatly cooled by the expansion of the universe. The light is detected as microwaves and millimeter waves, similar in wavelength to the radiation picked up by a television. In fact, about 1% of the static on a TV tuned between stations has come directly from the Big Bang.

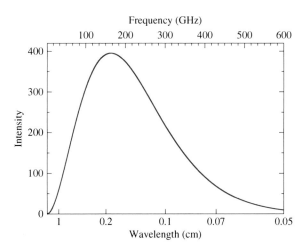

FIG. 7.1 The prediction of the Big Bang theory for the energy spectrum of the cosmic microwave background radiation compared to the observed energy spectrum. The FIRAS experiment on NASA's COBE satellite measured the spectrum at 34 equally spaced points along the blackbody curve. The error bars on the data points are smaller than the thickness of the line used to draw the curve

Penzias and Wilson made their measurement at a wavelength of about 7 cm. Soon after, Roll and Wilkinson made a measurement at 3.2 cm. The intensities of the radiation measured at these two wavelengths suggested a radiation temperature of about 3 K. The intensity measured by Roll and Wilkinson was higher than that found by Penzias and Wilson by just the right amount. Of course two measurements are not enough to confirm that the radiation follows a curve. The idea was then to make more measurements but how and at what wavelengths?

From the ground, one can in fact study the intensity of the radiation from wavelengths of about 1 m down to a third of a centimeter. The cosmic background radiation peaks at wavelengths shorter than this. Why not make measurements where the cosmic radiation intensity is highest? The problem is that the atmosphere becomes increasingly opaque at wavelengths shorter than 0.3 cm. For this reason, prior to 1989, the short wavelength part of the spectrum was studied using rockets and balloon experiments.

In addition to the problem of atmospheric absorption there is the fact that our galaxy is a strong emitter of radiation at both radio and infrared wavelengths. This foreground emission interferes with measurements of the cosmic background. We are really interested in light coming from behind the galaxy. It is as if you are trying to look at some faint star in the sky but the city lights make the night sky too bright. The nearby bright light makes it harder to see the distant faint light. At radio wavelengths the emission from our galaxy comes from electrons emitting radio waves by spiraling in magnetic fields. At infrared wavelengths the emission from our galaxy comes mostly from dust. There is a window at a wavelength of 0.4 cm between the long radio wavelengths and the shorter infrared wavelengths where the cosmic background radiation is dominant.

The brightness of the cosmic background radiation depends on the direction one is looking. These changes in intensity with small changes in direction reveal the first clumping of matter after the Big Bang, the seeds of galaxies. The cosmic background effectively carries with it an image of the universe when it was only 400,000 years old. To detect these galaxy seeds NASA decided to build a satellite. There was a strong motivation to detect the predicted brightness variations. Lack of detection would have meant that galaxies did not form by the gravitational collapse of slightly overdense regions in the early universe.

The Cosmic Background Explorer Satellite: NASA's Triumph

The COBE satellite discovered the primordial density fluctuations that formed the large structures such as the Sloan great wall of galaxies. When the results were presented at the annual meeting of the American Astronomical Society there was a standing ovation. Three instruments were on board the satellite; a spectrograph designed to measure the spectrum of the background to high accuracy, a radiometer designed to map the cosmic radiation, and an infrared experiment designed to search for the cosmic infrared background radiation.

The radiometer was designed to operate at wavelengths where the contamination from our galaxy is minimal. The radiometer detected the variation in the intensity of the background as shown in Fig. 7.2. The amount of variation was found to be at a level of one part in 100,000. Studies of the early universe moved from speculations to theories that could be tested with actual measurements. The whole field of cosmology changed.

The satellite orbited the Earth 14 times a day. The infrared instruments were kept cool using 650 l of liquid helium. These instruments had to be kept cool to prevent them from emitting infrared radiation. Twenty four hours worth of data could be transmitted to the ground and stored in 9 min. COBE was funded as part of NASA's Explorer mission program, it had a relatively small cost of 30 million dollars, less than we spend each morning in the war in Afghanistan. COBE triumphed because it held the only route to observing the cosmic background fluctuations.

How do we know that the background radiation is indeed of cosmological origin? Fred Hoyle and others used to argue that the radiation is produced by starlight scattered by iron needles. These scientists did not believe that a Big Bang took place, so they needed an alternative explanation for the existence of the microwave background. Such needles could be produced in the

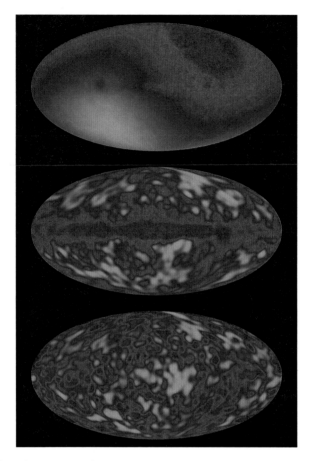

FIG. 7.2 The sky images from the COBE satellite differential microwave radiometer. Each map shows the information from the whole sky much as we can show a map of the spherical earth on a page. The *top* map shows all of the information. This map is dominated by an effect called the dipole, caused by the motion of our galaxy relative to the background radiation. In the *middle* map the dipole effect has been removed leaving the cosmic background emission and the emission from our Milky Way Galaxy. The emission from our galaxy is the broad horizontal strip in the middle. In the *bottom* image only the variations in brightness due to the cosmic background radiation remain. These variations are very small (Credit: NASA)

wake of supernova explosions, which are known to produce iron. There are two problems with this idea. First, it is very difficult even in the laboratory to produce blackbody radiation that follows the theoretical curve as precisely as the observed background does (Fig. 7.3). It is hard to imagine how iron needles would do the trick. They would have to be heated by starlight which is not at

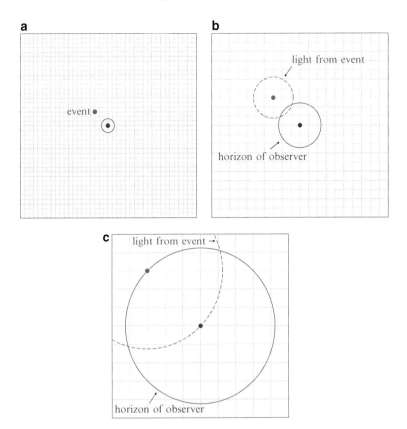

FIG. 7.3 This figure depicts space as a grid. Our location in the universe is depicted as the *red point*. The circle depicts our horizon which is how far into space we can see (*upper left panel*). The radius of the circle is determined by the time elapsed since the Big Bang. Shortly after the Big Bang this radius is relatively small. We have depicted an event, say a supernova going off at this time, as a *green dot*. The event is located three by three grid cells away. The panel **b** shows that after some time space has expanded and the light from the event has traveled outward in all directions (*green dashed circle*). When the light from the event finally reaches us as shown by the *green dashed line* reaching the *red dot* we observe the event with our telescopes (panel **c**)

a uniform temperature, nor homogeneously distributed in space. Needles produced in supernova explosions would be distributed like stars which are not at all homogeneously distributed around us. The Big Bang, as unlikely as it may seem, is the only plausible mechanism we have come up with to account for the background radiation.

A good test would be to measure the radiation temperature at high redshift. When the universe was younger the background radiation would have had a higher temperature. In 1994, astronomers used the Keck telescope to study absorption lines of carbon in gas clouds close to a quasar at redshift 1.8. One cloud has a measured temperature of 10.4 K while the other has a temperature of 7.4 K. The predicted temperature of the background radiation at redshift 1.8 is about 7.6 K. The measured temperatures have errors of about 1° or so. The agreement is fairly good but not perfect. The interpretation of the data is complicated by effects of molecular collisions in the clouds. The measurements do confirm that the temperature was higher in the past.

The Journey of Light Through the Expanding Universe

After neutral atoms form, the light from the Big Bang starts on its long journey through space until some of it is caught by our telescopes. Figure 7.3 shows three snapshots of light propagating in an expanding universe. The grid marks the location of points that are taking part in the expansion. In the upper left panel an event takes place such as a supernova explosion. Our location is marked by a red dot. The cosmic (or Hubble) expansion takes the red and green dots further from each other. The small red circle in the upper left panel shows our horizon; the part of the universe that we can see at any given time. Our horizon expands until it reaches the green dot in the third panel. At that time we can see the event, that is to say the pulse of light has finally reached us.

Figure 7.4 illustrates the fact that events that took place simultaneously at similar distances from us are observed simultaneously. It is a generalization of Fig. 7.3 to multiple events. In the two dimensional case shown in Fig. 7.4 events which took place at the same time in the past are located on a circle centered on our location. In the three dimensional world this circle is actually a sphere. The event for the cosmic background radiation is the point at which the universe became neutral and thus transparent and light could travel freely. The formation of hydrogen atoms occurred everywhere at the same time because it only depends on the temperature of the background radiation. When we observe the cosmic

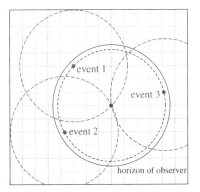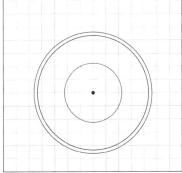

FIG. 7.4 *Left panel*: all events that happened simultaneously on the *blue dashed circle* are observed simultaneously. *Right panel*: the *green circle* shows the location in space of points we can observe today as they appeared about 8 billion years after the Big Bang (redshift 1). The *blue circle* shows the location of points that we can observe as they appeared 380,000 years after the Big Bang (the cosmic background) or a redshift of 1,100. The *red circle* shows the location of points that emitted light at the moment of the Big Bang

background radiation we are seeing in some sense a snapshot of the surface of a sphere surrounding us. The background radiation conveys direct information about the state of the universe about 400,000 years after the Big Bang. The surface that we see with the cosmic background is known as the surface of last scattering. We experience the same thing when we look at clouds in the sky. We cannot see inside the clouds but we can see the edge of the clouds where light is no longer scattered and can make its way directly to us (Fig. 7.5).

COBE's Successor: The Wilkinson Microwave Anisotropy Probe

The WMAP satellite was designed to carry out detailed measurements of the brightness variations found by COBE. By analogy with Fig. 7.5 it was as if we had discovered the existence of clouds high in the sky and we now wanted to learn about the nature of these clouds by mapping their structure in detail. As we shall see the detailed maps contain a wealth of information.

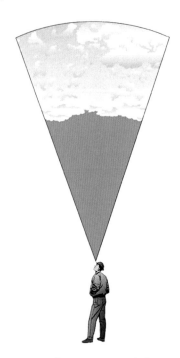

FIG. 7.5 We can compare our observations of the cosmic background radiation to observations of clouds. When we look up at clouds we see their surface where they become transparent but we cannot look further inside the cloud (Credit: NASA/WMAP Science Team)

We can associate a size with brightness measurements. It is as if we were measuring the population of country by laying a grid with a separation of 200 miles between grid lines on a map and counting the number of people in each square. This grid would reveal the most general pattern of population; more people on the coasts less in the Midwest. By moving the grid lines closer together, say 10 miles, we can get much more information revealing the presence of towns and cities, since some squares will have hundreds of thousands of people while others will be at zero. We can then estimate the variations in the numbers in the squares for a given square size. The variations will be largest for a grid size that corresponds to some real feature such as the size of a city.

This was the idea behind mapping the cosmic background radiation in more detail. The theory predicted that for certain scales (angles in the sky) the variations in brightness should be larger than others. In particular the prediction was that the

FIG. 7.6 The COBE satellite (*upper left*) made the first measurements of temperature differences in the cosmic microwave background. The BOOMERANG (*middle*) and MAXIMA balloon experiments mapped smaller portions of the sky but at much higher resolution. With the WMAP satellite's high-resolution map of the whole CMB sky (*lower right*), high precision measurements of the age, density, and curvature of the universe became possible for the first time (Credit: NASA, the National Science Foundation, and Lawrence Berkeley National Laboratory)

variations in brightness should be relatively large if one observed the sky with a 'grid size' of about 1° (about twice the diameter of the full Moon). Measurable effects were also predicted for smaller grid sizes.

The search for these effects was initially carried out using measurements made from balloons. They scanned small patches of the sky but with greater angular detail than could be seen by COBE. The balloon data showed the presence of the main predicted feature on a scale of 1° but also the presence of a second and third one on scales smaller than 1°.

In 2001 the WMAP satellite was launched. It measured the entire cosmic background sky at five frequencies. Figure 7.6 illustrates how WMAP revealed the background radiation in much more detail than COBE did. WMAP detected features as small as one fifth of a degree in the sky (about one third of a moon diameter). This is about 30 times more detail than could be seen with COBE.

Scientists used the WMAP data to measure several numbers of fundamental significance to cosmology; the curvature of space, the baryon (atomic) density, the dark matter density and the dark energy density. All these numbers have been measured to an unprecedented accuracy of 1% or better. WMAP ushered in the age of precision cosmology. How then was this detailed information extracted from the WMAP data?

The Music of the Big Bang

Photons that find themselves in a high density region when the universe becomes transparent will have to climb out of that region to reach us. Just as it costs a rocket energy to escape from the Earth's gravity, it costs a photon energy to escape the gravitational field of a dense region of the universe. The result is that the photon acquires an initial redshift right at the beginning of its journey towards our telescopes. This photon, in comparison with a photon that started its journey from a place of average density, will have a slightly longer wavelength by the time it reaches us. This gravitational redshift is a small addition to the dominant effect which is the cosmological redshift due to the expansion of the universe.

When matter is still in the form of free electrons and protons, dark matter can compress baryons and radiation from one place to the next, resulting in temperature variations at a redshift. This is because the baryons feel the gravity of a dark matter clump and start falling towards the center of the clump but they feel a pressure from the photons that counters the force of gravity and pushes the matter outwards. What results is an oscillation, a bouncing back and forth, not dissimilar to a weight on a spring, that can bounce up and down. So some of the photons are compressed and at slightly higher temperature and some are at slightly lower temperature depending on whether the photons have over-expanded. When the universe turns neutral, the effect stops because photons then travel freely through space.

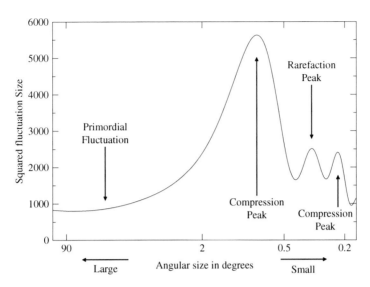

FIG. 7.7 The size of brightness variations in the cosmic background radiation as measured by the WMAP satellite. The largest variations occur at a size close to a degree, twice the size of the full Moon. This is because this is where the plasma waves have their largest effect on the radiation. Also at scales (angles) smaller than this fluctuations get washed out (see Fig. 7.9)

Clumps of different size oscillate at different speeds in a plasma. The strongest observable effect comes from large clumps in which the plasma has just had time to be compressed once before neutral atoms form. For smaller clumps the plasma can get compressed and bounce back. Maps of the cosmic background contain the frozen imprint of these plasma sound waves. In this sense we can see the music of the Big Bang.

How do we see these sound waves? We measure the brightness variations of the background radiation over the whole sky at small and large angles. This is like measuring the height of small wavelength and large wavelength waves on the ocean. To do this we smooth the radiation map (Fig. 7.6) and then measure the brightness variations. The smoothing erases all the brightness variations smaller than the smoothing size. We can do this from large angles (up to 90°) down to quite small angles 0.2°. For reference the full Moon diameter is about 0.5°.

Figure 7.7 illustrates this. We can theoretically predict the physical size of the first peak. The angular size in the sky of an object of known size and known redshift depends on the geometry

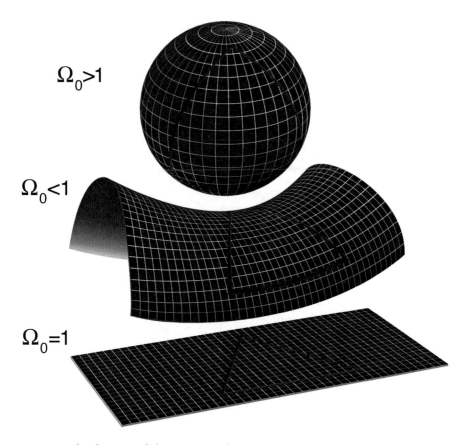

$\Omega_0 > 1$

$\Omega_0 < 1$

$\Omega_0 = 1$

FIG. 7.8 The density of the universe determines its geometry. If the density of the universe exceeds the critical density, then the geometry of space is closed and positively curved like the surface of a sphere. We use the number Ω to characterize the density of the universe measured in the units of a critical density. If the density of the universe is less than the critical density, then the geometry of space is open, negatively curved like the surface of a saddle. If the density of the universe exactly equals the critical density, then the geometry of the universe is flat like a sheet of paper. The WMAP measurements of the angular size of the first peak in Fig. 7.7 show that we live in a flat $\Omega = 1$ universe (Credit: NASA/WMAP Science Team)

of space. The conclusion is that the geometry of space is flat like the surface of a table, rather than curved like the surface of a sphere such as the Earth. If the curvature of space had been found to be positive like a sphere, the angular size of the first peak in Fig. 7.7 would have been found to be larger, on the other hand if the curvature had been negative like the middle figure in Fig. 7.8 the angular size would have been observed to be smaller. The fact

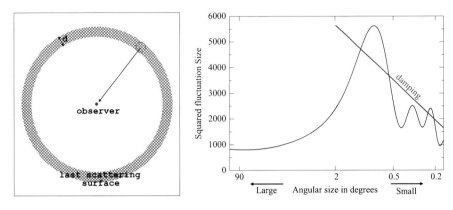

FIG. 7.9 Recombination does not happen instantaneously so the last scattering surface has a finite thickness (labeled d in the figure). As a result of this, fluctuations (variations in temperature) on scales smaller than d are washed out which causes us to observe smaller variations in brightness on the sky at small angle separations. This effect is known as diffusion damping. The effect on the measured spatial variations in the background intensity is shown by the *blue line* in the figure on the *right*

that the universe has a flat geometry means we can combine the cosmic background and supernova data to determine the relative percentage of dark energy and dark matter in the universe. It turns out that dark energy accounts for 72% of the energy density of the universe. A geometrically flat universe can recollapse eventually or expand forever. The data indicate that our universe will expand forever and that the expansion rate will accelerate.

The formation of neutral atoms did not occur instantaneously. To use our cloud analogy from Fig. 7.5, we can see just a little bit inside the cloud because it does not instantly become opaque like a price of wood and we can indeed see a little ways inside the surface of last scattering. This has the consequence that the smaller angled variations in brightness get smoothed out as illustrated in Fig. 7.9. This explains why the first peak is so much more pronounced than the second and third ones.

The Parameters of the Universe

WMAP tells us that the universe has flat geometry. The supernova data prove the existence of dark energy. By combining these two results with galaxy data we can calculate with good precision the

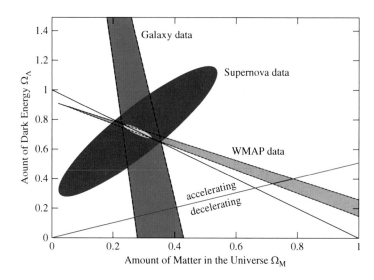

FIG. 7.10 A plot of dark energy density Ω_Λ versus matter density Ω_M. The *shaded areas* represent areas of the plot that are consistent with certain observations. The *blue ellipse* shows the constraints from the Type Ia supernova data. The values of Ω_Λ and Ω_M must lie within the *blue area* to agree with the supernova data. The values must also lie within the *green shaded area* to agree with the observations of the clustering of galaxies. The values must lie within the *orange area* to agree with the WMAP cosmic background observations. The values must lie within the *grey ellipse* to agree with all three sets of data. The fact that the *grey ellipse* lies above the *red line* means the expansion of the universe is accelerating and the universe will expand forever. The fact that the *black line* goes through the *grey ellipse* means the geometry of the universe is flat (see also Fig. 7.8)

relative amounts of matter and dark energy in the universe. We do not at this time understand the origin of dark energy or the nature of the dark matter. The three sets of measurements (WMAP, supernovae and galaxies) each enable us to roughly estimate these parameters but when we *combine* all three measurements the range of possible values is narrowed down considerably. Figure 7.10 illustrates this. The blue ellipse represents the range of values allowed by the supernova data. The orange area represents the region allowed by the WMAP data. The orange area lies quite close to the black line which represents a flat universe as we explained earlier. In fact the overlap between the supernova data constraints (blue ellipse) and the WMAP data (orange triangle) leaves the smaller grey region. This region corresponds to a flat universe. The grey ellipse lies in a part of the diagram well above the red line that separates an accelerating expansion rate from a decelerating expansion rate. The geometry of the universe is flat and the universe is expanding at an increasing rate due to the presence of a mysterious dark energy that currently account for 72% of the density of the universe, the remaining 28% consisting of cold dark matter and baryons. The baryons only account for 4.6% of the density of the universe today. The age of the universe according to WMAP data is 13.73 billion years. These numbers characterize the weight, shape and fate of the universe.

The WMAP measurements have enabled us to determine the parameters of the universe to high accuracy but they do much more. To understand how galaxies formed we must know what the universe looked like before they formed. If we compare the density of our galaxy with the density of the universe we find that our galaxy is today about 200,000 times denser than the mean density of the universe. The measurements of very small temperature fluctuations in the background radiation temperature. These variations imply density variations of one part in 100,000. That is to say the density of the universe varied little from one place to the next when the universe was very young. You might measure the density to be 1.0000 times the mean density in one place and 1.00001 times the mean density in another. This is very different from the universe today where the density varies a lot from one place to the next. As we have mentioned, the density in our galaxy is many times larger than the mean density of the universe today.

The cosmic background measurements made it possible to extract physical parameters from the galaxy surveys. The brightness variations gave us a benchmark for starting the computer simulations that explain the appearance of the universe today. This sets the stage for Part III of this book which is about the attempt to see and measure the formation of galaxies during the time period between the galaxy maps and the cosmic background observations. This is the new frontier of cosmology.

Further Reading

The Music of the Big Bang: The cosmic Microwave Background and the New Cosmology. A. Balbi, Berlin, Springer-Verlag, 2010.

The 4% Universe. R. Panek, Boston, Houghton Mifflin Harcourt, 2011.

Finding the Big Bang. P. Peebles, L. Page and R. Partridge, Cambridge, Cambridge University Press, 2009.

Observational Cosmology. S. Serjeant, Cambridge, Cambridge University Press, 2010.

Part III
The Search for the Cosmic Dawn

In this final part of the book we explore the frontier of cosmology; the search for the first stars, galaxies and black holes. When the universe cooled down sufficiently for neutral hydrogen atoms to form it entered a cosmic dark age, during which there were no planets, stars, or galaxies, just warm hydrogen and helium gas. Chapter 8 describes this dark age and the plans underway to detect the hydrogen gas as it cooled and was concentrated into clumps by gravity, forming the first stars. Astronomers believe that these primordial stars were hundreds of times larger than our Sun and extraordinarily hot, bright, and short-lived. Their intense ultraviolet radiation broke apart neutral atoms, initiating a longer, more gradual epoch of reionization. The light produced by these stars is sometimes referred to as the cosmic dawn.

We describe the search for this cosmic dawn in the last four chapters of this book. Chapter 9 is an overview of high redshift galaxy studies. Chapter 10 explores the idea that the first galaxies to form after the big bang have survived to the present day as faint fossil relics that may be detectable close to our Milky Way galaxy. The next generation of telescopes that will observe the cosmic dawn are previewed in Chap. 11. Chapter 12 introduces the successor to the Hubble Space Telescope, the James Webb Space Telescope and its planned observations of the cosmic dawn.

8. The Search for Light in the Dark Ages

Following the recombination and the formation of the first atoms, the early universe was a nearly formless primordial soup of dark matter and gas: there were no galaxies, stars, or planets. This was truly the dark ages. Things began to change when the slightly denser regions began to contract under the relentless pull of gravity. It took a few hundred million years, but eventually these dense regions gave birth to first stars, and black holes so that the universe became filled with light. These events lie largely in the realm of theory today, and existing telescopes can barely probe this mysterious era. Over the next decade, we expect this to change.

Committee for a Decadal Survey of Astronomy and Astrophysics

From half a million years to five hundred million years after the Big Bang, the hydrogen in the universe was in the form of neutral atoms. At earlier times, the cosmic background radiation had sufficient energy to keep the hydrogen atoms ionized. At later times the ultraviolet radiation from young stars and quasars was sufficient to keep the hydrogen gas between the galaxies ionized. The period between recombination and reionization during which the hydrogen gas was in the form of neutral atoms is known as the dark ages. The light emitted at 21-cm during the dark ages will be redshifted to meter wavelengths. Radio telescopes have been built to detect the imprint of neutral hydrogen at these early times. The majority of light in the universe during this period was in the form of cosmic background radiation left over from the Big Bang. During this period this radiation travels (almost) unimpeded through space. We say almost because the neutral hydrogen will leave a small imprint on the light by either adding or removing light at specific wavelengths. The physical processes that control

G. Rhee, *Cosmic Dawn: The Search for the First Stars and Galaxies*,
Astronomers' Universe, DOI 10.1007/978-1-4614-7813-3_8,
© Springer Science+Business Media, LLC 2013

the temperature of the hydrogen gas vary with time resulting in the fact that at some redshifts the gas could be detected in emission and at others in absorption.

Observations at meter wavelengths are complicated by the emission from the Milky Way galaxy and also by radio and cell phone emissions on earth. This foreground contamination by our galaxy has to be removed from the data before the cosmic background radiation and associated small distortions can be detected.

A large research effort is going into the design and construction of radio telescopes to detect this faint glow of light from the dark ages. This presents a huge technical challenge but the rewards will be commensurate with the effort. How then can we hope to observe the dark ages, the period between recombination and reionization?

How to Detect Atomic Hydrogen

The most common element in the universe, hydrogen, consists of one proton and one electron. For our present purposes we can think of the atom as a very small solar system with one planet (the electron) orbiting the Sun (proton). In the atom, the electron is only allowed to be in certain orbits or energy states as we call them. The electron can move from one orbit to the next be emitting or absorbing light of known color (wavelength) and energy. Like a planet that spins on its axis, the electron also has spin, but it can only have one of two spin values in the atom, spin up or spin down (Fig. 8.1). The electron can emit light by changing its spin from up to down and absorb light by changing its spin from down to up. The wavelength of light that is absorbed or emitted during a spin change is 21-cm corresponding to a frequency of 1,428 MHz. This wavelength is located in the radio part of the electromagnetic spectrum. This transition was predicted to be observable by the Dutch astrophysicist Henk van de Hulst. The 21-cm emission from our galaxy was first detected by Ewen and Purcell at Harvard in 1951. Neutral hydrogen gas thus reveals itself to us through the emission of radio waves. Light emitted by neutral hydrogen gas in the early universe will be redshifted to meter wavelengths by the time it reaches us.

Our current calculations and WMAP observations suggest that reionization began at a redshift of about eleven and was over

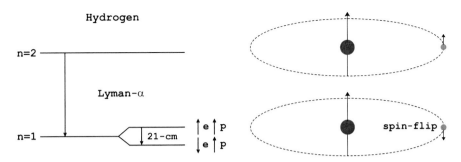

FIG. 8.1 Transitions of the hydrogen atom. The 21-cm transition is between two states of slightly different energy where the spin of the electron (e) is aligned with that of the proton (p) and where the spins are opposed. A spin flip of the electron results in the emission of a photon with wavelength 21-cm. The transition between electron orbits (energy levels 1 and 2) results in the emission of a much more energetic photon in the ultraviolet part of the spectrum with wavelength 1,216 Å or 1.216×10^{-5} cm

by redshift seven. For 21-cm these two redshifts correspond to observed wavelengths of about 2.5 and 1.7 m respectively. The Square Kilometer Array telescope in Australia is designed to detect neutral hydrogen at these redshifts.

Observing the Spectrum of the Cosmic Dawn

We have specified the redshift range of interest, and thus the wavelength range to which the 21-cm light will be redshifted. But what do we expect the signal to look like? To answer this question we must examine the physical picture of the universe during the dark ages. The universe at this time consists of a gas of mostly hydrogen and some helium. We also have the redshifted cosmic background radiation as well as dark matter assembling into clumps (or halos) of various masses.

The hydrogen gas modifies the spectrum of the cosmic background radiation in ways that should be detectable with modern radio telescope technology. The specific frequencies at which radio emission peaks and troughs should occur are predicted by our theories. The detection of such features in the spectrum of the cosmic background radiation would place strong constraints on the nature of star and galaxy formation in the early universe.

It is conceivable that there is no visible effect. This would mean that on average the hydrogen gas absorbs as many photons from the background radiation as it emits. The second possibility is that the gas is at a hotter temperature than the background radiation. In that case the gas would produce an excess of 21-cm radiation over what is already present from the cosmic background. The third option is that there are too many electrons in the low energy state (spin down). In that case the hydrogen gas would absorb a net amount of energy from the background radiation and we see an absorption feature in the spectrum. Which of these is the correct option? According to theoretical calculations, it depends on the wavelength that we use to carry out the observations.

The effects are illustrated in Fig. 8.2. When the blue line is above zero we expect the gas to be seen in emission, when the line is below zero the gas absorbs more light than it emits and photons are removed from the background.

When the universe is younger than 10 million years old we expect no visible effect on the background radiation because the gas has the same temperature as the cosmic background radiation. At later times the gas cools faster than the radiation as the universe expands. The result of this is that the gas is seen in absorption. Eventually the gas dilutes so much that it no longer produces detectable effects on the radiation. After the first stars form, they emit ultraviolet radiation which heats up the hydrogen gas, with the effect that the gas emits an excess of photons and can be seen in emission down to redshifts of about ten. By redshift six the universe is almost completely reionized. This puts an end to the absorption and emission of 21-cm radiation by hydrogen atoms since there are very few neutral atoms left.

We have to look through the Milky Way galaxy in order to see what is happening at high redshifts since our solar system is located inside the Milky Way. Our galaxy is a strong emitter of radio waves at all wavelengths (Fig. 8.3). In fact the Milky Way galaxy is 1,000 times brighter than the radio signal we are trying to detect. To have any chance of detecting the signature of the high redshift hydrogen emission we have to very accurately model and subtract out the Milky Way emission. Techniques are being developed to subtract the foreground emission based on the fact that it varies smoothly with wavelength and has a different spatial distribution to the redshifted neutral hydrogen emission.

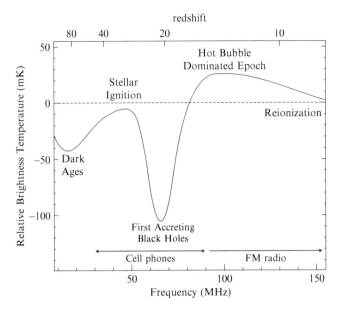

FIG. 8.2 The *blue line* shows the average spin-flip signal in a simple model of the dark ages and the reionization era. The horizontal axis gives the observed frequency. Because we are talking about the 21-cm line, the observed frequency is a measure of the redshift at which the 21-cm radiation was emitted. The brightness temperature plotted on the vertical axis is a measure of the excess or deficit of light measured relative to the cosmic background radiation intensity. We labeled four points towards the end of the dark ages. Stellar ignition marks the point at which we expect the first stars to appear. The second point around redshift twenty marks the period when the first black holes begin to heat the hydrogen gas. The hot bubble epoch marks the beginning of reionization and we see the end of reionization when all the hydrogen gas is ionized (Credit: Burns et al. Advances in Space Research, 49, 18 2012, reprinted with permission by Elsevier)

Images of the Dark Ages

Throughout the dark ages the neutral hydrogen is following the dark matter and getting more clustered with time. We do not have observational evidence of this. Images of the neutral hydrogen gas over a range of redshifts would map the evolution of dark matter during the dark ages. The hydrogen gas distribution yields direct information on the dark matter density variations that were generated a very small fraction of a second after the Big Bang occurred.

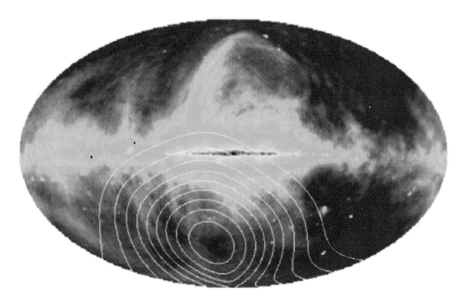

FIG. 8.3 Radio map of the sky at 100 MHz corresponding to redshift 13 for the 21-cm wavelength of hydrogen or about 350 million years after the Big Bang. The emission seen in this map comes mostly from the Milky Way galaxy. The plane of the edge on disk of our galaxy is seen as the horizontal feature in the center of this image. The emission from our galaxy has to be modeled and subtracted with very high precision if we can hope to see the redshifted 21-cm emission from the dark ages. The *white lines* represent the part of the sky that a radio telescope orbiting in space would look at credit (Credit: Burns et al., Advances in Space Research, 49, 18 2012; derived from a model by de Oliveira-Costa and collaborators, reprinted with permission by Elsevier)

At redshifts smaller than about 25, the neutral hydrogen observations reveal something else about our universe. We believe that around this redshift the first stars and galaxies form and the universe starts to get reionized. The ultraviolet light emitted by the young stars and galaxies forms holes of ionized gas in the neutral hydrogen. The neutral hydrogen distribution starts to look like swiss cheese, at first Gruyère and later Emmental (i.e. the holes get bigger).

The 'holes' in the neutral hydrogen are located around the first stars (Fig. 8.4). A hole is not an absence of gas but an absence of neutral gas (atoms); the gas is present in the form of free electrons and protons. The holes or bubbles near the end of reionization are believed to be about 100 million light years in size. Such bubbles would appear to us on the sky to be about the size of the full moon (as we see it from earth). After reionization is complete,

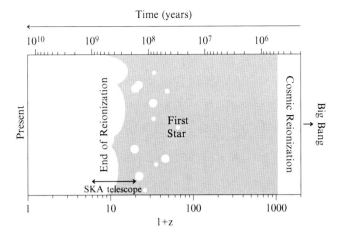

FIG. 8.4 Some key points about the reionization history of the universe are illustrated above. The age of the universe in years is shown on the *top* horizontal axis and the corresponding redshift is shown on the *bottom* $(1 + z = 1$ is the present day). The *white areas* represent parts of the universe that are ionized and the *gray region* illustrates the part that is neutral. The *small white circles* represent the ionization regions around the first stars. These stars form in dark matter halos that have about 100,000 times the mass of the Sun. The 21-cm line observations will probe the era of the universe represented by the *grey background* (Credit: based on a figure by R. Barkana and A. Loeb)

neutral hydrogen is only present inside galaxies. The dark matter distribution can then be traced by the clustering of galaxies where each galaxy is treated as a point (see Chap. 5).

To recap; prior to a redshift of 25, the neutral hydrogen traces the dark matter distribution. Using this technique we can potentially map a much larger volume of space than the cosmic background radiation snapshot does. We can map the dark matter fluctuations down to much smaller sizes than those measured by the cosmic background. Between a redshift of 25 and a redshift of 6, ionized hydrogen bubbles of increasing size form until the bubbles merge and the neutral hydrogen between galaxies is gone. At redshifts less than six the neutral hydrogen maps the location of galaxies which are themselves tracers of the underlying dark matter distribution. The potential wealth of redshifted 21-cm observations has motivated radio astronomers to build meter-wave radio telescopes suited to high redshift neutral hydrogen studies.

Observing the Dark Ages with Radio Telescopes

One goal is to collect enough light to get a spectrum of the redshifted 21-cm sky. Spectroscopy is in this sense easier than imaging since we do not need to map the sky. We expect the light to be coming from all directions (like the cosmic background radiation) so a large part of the sky can be studied at one go. One example of this approach is the Experiment to Detect the Global EOR Signature (EDGES). The EOR in the acronym stands for End of Reionization. This experiment is deployed at the Murchison Radio Astronomy Observatory in Western Australia to measure the radio spectrum between 100 MHz (3 m) and 200 MHz (1.5 m). The design is simple; an antenna, an amplifier, and a computer are connected to a solar energy source. The EDGES experiment has produced the best broad band radio spectrum of the sky and has ruled out instantaneous reionization.

Proposals have also been made to carry out similar measurements from space. One could place a satellite in orbit around the Moon and make measurements from the far side of the Moon. This method eliminates the most intense man-made foreground emission such as radio transmissions from radar, radio and TV stations.

The hope is that we will detect events such as the formation of the first stars $(30 > z > 22)$. The first black holes in the universe may have formed in the redshift range $22 > z > 13$ as remnants of the first stars. The effect of these black holes would be to transform the 21-cm signal from absorption into emission. As most of the gas becomes ionized $(13 > z > 6)$ the signal from neutral hydrogen gas is no longer detectable. The spectral information can potentially tell us when the first stars formed in the universe, when the first black holes formed, when reionization began and when it ended.

To image these events we need to use radio interferometers. An interferometer combines the signal of many individual radio receivers. Radio telescopes being built for this purpose include the low frequency array (LOFAR) in the Netherlands which has 44,000 antennas covering an area of 1,500 km diameter (most of the antennas will be in the Netherlands but a few will be placed in

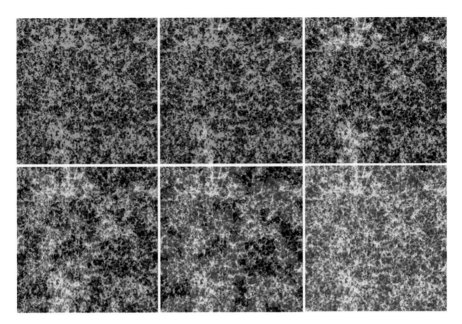

FIG. 8.5 Slices through a model universe simulating the reionization process. The neutral hydrogen is shown in *green*, the ionized hydrogen in *orange*. The redshifts of each slice in the *top row* from *left* to *right* are 18.5, 16.1 and 14.5. The redshifts of each slice in the *top bottom* from *left* to *right* are 13.6, 12.6 and 11.3. This covers a period from 250 million years after the Big Bang to 400 million years after the Big Bang. The angular size of this box viewed on the sky would be roughly one and half times the size of the full moon. By a redshift of 11.3 less than 1 % of the neutral hydrogen remains. Reionization proceeds through the overlap of ionized bubbles (Credit: I. T. Iliev, G. Mellema, U.-L. Pen et al. Simulating cosmic reionization at large scales I. The geometry of reionization, Monthly Notices of the Royal Astronomical Society, (2006), 369, 1625–1638, by permission of Oxford University Press on behalf of the Royal Astronomical Society)

France, Germany, the UK and Sweden). The Murchison wide field array located in the radio quiet western Australian outback will consist of 2,048 antennas with separation of up to 3 km.

Figure 8.5 shows what one might expect the redshifted 21-cm line sky to look like. The images cover an area of the sky comparable to the full moon. The redshift range is 18.5 to 11. At first the sky is covered by neutral hydrogen seen in emission (shown in green in the figure). Gradually the neutral hydrogen (green) gets converted to ionized hydrogen (shown in orange). At a redshift of eleven (bottom right panel) there is almost no detectable neutral hydrogen.

The instrument best suited to carrying out these mapping measurements is the Square Kilometer Array (SKA) to be located in South Africa and Australia. The SKA will search for neutral hydrogen in the redshift range from 20 to 7, precisely where the reionization transition is expected to take place (Fig. 8.4). The total collecting area of the SKA will be $1\,km^2$. To achieve this, the SKA will use 3,000 dish antennas, each about 15 m wide as well as two other types of radio wave receptor, known as aperture array antennas. The antennas will be arranged in five spiral arms extending to distances of at least 3,000 km from the center of the array. The central regions in Australia and in South Africa will contain cores each 5 km in diameter; one for each antenna type. The aperture array antennas will extend to about 200 km from the core regions. In Southern Africa the dishes will be positioned in distant stations out to at least 3,000 km.

The SKA will be a factor of ten more sensitive than the currently most sensitive instrument, the Expanded Very Large Array radio interferometer. The first observations will be carried out in 2019 and the full array will be operational by 2024. New computer systems will be necessary to deal with the vast amount of data produced by SKA. The radio antennas will produce a flow of data of up to 10^{15} bits per second, which is 100 times the total amount of internet traffic today.

We have already mentioned that part of the trouble in making low frequency measurements from the ground lies in what is called radio frequency interference from man-made sources such as television transmission and cell phones. The FM band used for broadcasting radio goes from 90 to 1,100 MHz which is precisely where the effects of reionization are strongest. The radio frequency range from 30 to 110 MHz has been almost exclusively allocated to cell phone usage and TV and radio. This makes it necessary to locate the telescopes in parts of the world, such as the deserts of Australia, that are radio quiet.

The study of the dark ages is a field that holds great promise. Astrophysicists are working hard to predict the properties of neutral hydrogen prior to and during the epoch of reionization. We hope to answer several questions. When did reionization take place? How long did it last? What sources were responsible? The hope is that in the next decade we will go from the realm of speculation to anchoring our theoretical ideas on sound physical

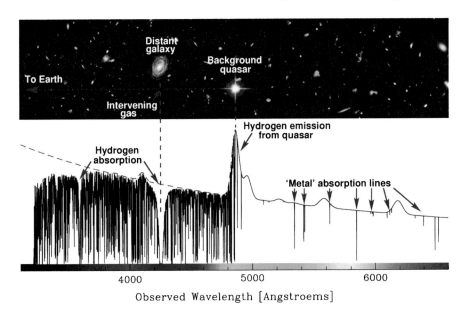

FIG. 8.6 Quasar spectra tell us about the gas intervening between us and the quasar because hydrogen gas removes light from the quasar spectrum at known wavelengths. These wavelengths get shifted towards the *red* telling us about the distance from us of the hydrogen gas clouds. The *solid red line* marks the quasar spectrum at wavelengths longer than the peak of emission. The *dashed line* shows what we think the spectrum would look like in the absence of hydrogen absorption (Credit: John K. Webb, University of New South Wales)

measurements. At redshifts less than seven we can already observe neutral hydrogen in the universe using the spectra of quasars.

Neutral Hydrogen at z < 7: Evidence from Quasar Spectra

Optical spectra of quasars provide evidence that neutral hydrogen exists at high redshifts. The electron in a hydrogen atom can go from one orbit to another by absorbing and emitting light at certain known wavelengths. When hydrogen gas is cold, the electrons are all in the lowest orbit. The electron can jump to the next energy level by absorbing ultraviolet light with a wavelength of 1,216Å. If the gas is hot (but not ionized) the gas will emit light at that wavelength. We see this light as a feature in quasar spectra, it is referred to as the Lyman-α line (Figs. 8.6 and 8.7).

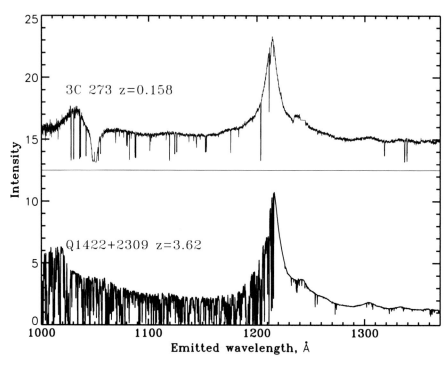

FIG. 8.7 The panel compares the spectra of two quasars at very different redshifts. 3C 273 is at a redshift of 0.158 whereas 1,422 + 2,309 is at a redshift of 3.62. The spectra are shown at their emitted wavelengths. The strong peak is the Lyman-α emission line. It is visible in both spectra. In the high redshift quasar spectrum the emission blueward (to the *left*) of the peak is lessened by the presence of many narrow *absorption lines* of neutral hydrogen known as the Lyman-α forest (Credit: Michael Rauch, Carnegie Institution of Washington, Sally Heap, Space Telescope Science Insitute, NASA/ESA, Bill Keel, University of Alabama)

Quasars are among the most luminous sources known to us. The light is emitted by gas spiraling into a black hole at the center of a galaxy. What we see as a quasar is the compact region in the center of a massive galaxy surrounding its central black hole. More than 200,000 quasars are now known. These were mostly discovered by the Sloan Digital Sky Survey. The most distant quasars have redshifts of about seven so their light was emitted less than 1 billion years after the Big Bang.

Figure 8.6 illustrates how the spectrum of a quasar reveals the presence of intervening hydrogen gas. Figure 8.7 shows spectra of two quasars at different redshifts. The spectra are shown in their rest frame so that the Lyman-α emission is at the same location on the x-axis.

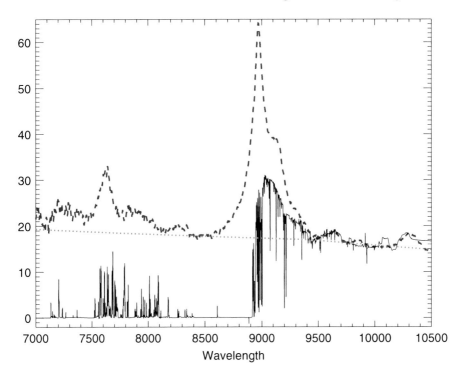

FIG. 8.8 Comparison of the observed spectrum of a quasar at redshift 6.37 (*black line*) with the expected spectrum based on low redshift quasar observations (*dashed red line*). To the left of the Lyman-α peak there is essentially no emission suggesting that there is sufficient neutral hydrogen present to entirely absorb the light from the quasar. Note however that it does not take much neutral hydrogen to produce this effect (Credit: Richard White, Robert Becker, Xiahui Fan and Michael Strauss, 2003, Astrophysical Journal, 126, 1, reproduced by permission of the American Astronomical Society)

At wavelengths longer than 1,216 Å the spectra look somewhat similar but at shorter wavelengths they look completely different. The higher redshift quasar (lower panel) shows many sharp dips. These dips are collectively known as the Lyman-α forest. Each dip is a feature known as an absorption line that is believed to be caused by a cloud of neutral hydrogen gas that lies between us and the quasar. Comparing these two spectra tells us that in the past there were many more such clouds than in more recent times. By comparing spectra of quasars at different redshifts we can probe the distribution of neutral hydrogen at various epochs during the last 13 billion years. As we go to redshifts larger than six something dramatic happens. Figure 8.8

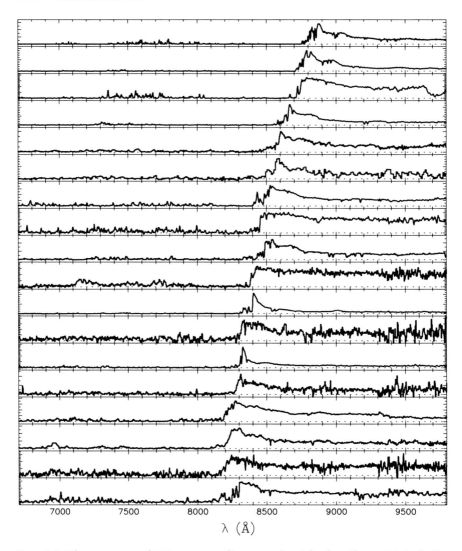

FIG. 8.9 The spectra of 18 quasars discovered with the Sloan Digital Sky Survey. The spectra are ranked from *bottom* to *top* in order of increasing redshift from a redshift of 5.74 for the *bottom* spectrum to a redshift of 6.42 for the *top* spectrum. The wavelength is plotted on the horizontal axis and the intensity on the vertical axis. As we go to higher redshifts there is less and less light remaining to blueward (to the left) of the Lyman-α emission line. In the quasars near the *top* of the figure almost all the light has been absorbed by neutral hydrogen gas located between us and the quasar. What we are seeing is a transition from an opaque universe (redshifts greater than six) to a transparent universe at lower redshifts, reflecting the end of the dark ages. For comparison see the quasar spectra in Fig. 8.7 at much lower redshifts of 0.16 and 3.6 (Credit: Fan et al. 2006, Astronomical Journal, 132, 117, reproduced by permission of the American Astronomical Society)

illustrates what the intrinsic spectrum of the quasar would look like if there were no intervening neutral hydrogen and what it actually looks like. The abundance of neutral hydrogen is high enough that the light blueward of Lyman-α has almost entirely been absorbed.

Figure 8.9 shows the spectra of 18 quasars in the redshift range $5.7 < z < 6.4$. The quasar spectra show that the neutral hydrogen fraction in the universe is changing with time. It is changing in the way that we expect with more neutral hydrogen present in the distant past. The data support the idea that the epoch of reionization is over by redshift six. It is not clear at present what objects were responsible for reionizing the universe. There are not enough quasars at redshift six to do the job. It is possible that star forming galaxies are responsible since young massive stars produce alot of ultraviolet radiation. The galaxies that we have detected at high redshifts are also not sufficient to do the job. Maybe an as yet undetected population of high redshift dwarf galaxies could do it. This is plausible because in the dark matter picture of galaxy formation the smaller galaxy halos form first.

Review

It may not be possible for us to directly detect the very first stars and galaxies to form in the universe. Our knowledge of their formation will then depend on the effect that they have on their surroundings. In that case, mapping and spectroscopy of neutral hydrogen at high redshift will be the only way to infer that the first generation of stars are forming. We have learned how the first stars will modify the spectrum of the cosmic background radiation that we measure at meter wavelengths. We have seen what an image of the neutral hydrogen sky might look like at the various wavelengths covered by the epoch of reionization. We learned how the quasars provide beams of light that can be used to map the neutral hydrogen clouds that survived the reionization of the universe. We measure the size of these clouds, how much neutral hydrogen they contain and their abundance of elements other than hydrogen. We can also study the clustering of these clouds which contains information on the dark matter distribution in the universe.

It would be most exciting if we could directly detect the earliest galaxies. We turn in the next chapter to the story of our attempts to push back to earlier and earlier times our observations of stars, galaxies and black holes.

Further Reading

The Reionization of Cosmic Hydrogen by the First Galaxies. Abraham Loeb in Adventures in Cosmology. David Goodstein, ed., Singapore, World Scientific. 2012.

Reionizing the Universe with the First Sources of Light. Steven Furlanetto in Adventures in Cosmology. David Goodstein, ed., Singapore, World Scientific. 2012.

How did the First Stars and Galaxies Form? Abraham Loeb. Princeton. Princeton University Press. 2010.

Observational Cosmology (Chapter 8, The Intervening Universe). Stephen Serjeant. Cambridge. Cambridge University Press. 2010.

9. Observing the First Galaxies

With increasing distance our knowledge fades, and fades rapidly. Eventually we reach the dim boundary - the utmost limits of our telescopes. There, we measure shadows, and we search among ghostly errors of measurement for landmarks that are scarcely more substantial.

Edwin Hubble

Astronomers today are close to seeing galaxies as they appeared just after the formation of the first stars. In this chapter we present three techniques currently used to find the most distant galaxies. The first two reveal galaxies that are forming stars at several hundred times the rate of the Milky Way galaxy. The third method selects galaxies with stars older than a billion years. We then discuss two accounting problems. We can measure the star formation rate over a large part of the history of the universe, but can we reconcile this birth rate with the number of star we actually see today? Secondly we know that sources of ultraviolet radiation kept the universe ionized at redshifts larger than six. We know that at lower redshifts the radiation from quasars is sufficient to do the job. At redshifts larger than six we do not see enough quasars to do the job but could galaxies be responsible?

Astronomy and Geology

We are getting to the point where we can observe the end of the epoch of reionization described in the last chapter. About 30 years ago our telescopes could study galaxies out to a redshift of one. We see these galaxies as they appeared about 6 billion years after the Big Bang. After the launch of the Hubble Space Telescope, it was realized that by making long exposure images of the sky one

G. Rhee, *Cosmic Dawn: The Search for the First Stars and Galaxies*, Astronomers' Universe, DOI 10.1007/978-1-4614-7813-3_9, © Springer Science+Business Media, LLC 2013

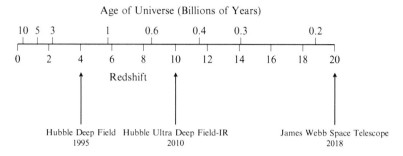

FIG. 9.1 As we probe the distant universe we observe galaxies with increasingly large redshifts. Each redshift value corresponds to a time after the Big Bang at which the light was emitted as shown by the upper axis labels in the figure. The most distant galaxies are seen in very long exposures taken with cameras on the Hubble Space Telescope. The Hubble Space Telescope imaged a very small region of the sky repeatedly for 35 h in 1995. The the experiment was repeated with a new camera on Hubble, with 55 h exposures in 2004. The 1995 experiment is called the Hubble Deep Field. The 2004 measurements are known as the Hubble Ultra Deep Field (Fig. 9.6). In 2010 an infrared version of the Hubble Ultra Deep Field imaged galaxies that may have redshifts as high as ten. The goal of the James Webb Telescope is to extend this work out to redshifts greater than 20. Each redshift value corresponds to a time after the Big Bang at which the light was emitted as shown by the upper axis labels in the figure (Credit: Rychard Bouwens)

could see much further. The famous image known as the Hubble Deep Field pushed the redshifts at which we could image galaxies to about four. We see these galaxies as they were 1.5 billion years after the Big Bang. In 2004, the Hubble Ultra Deep Field made it possible to look back over 13 billion years of the history of the universe to see galaxies as the appeared only 700 million years after the Big Bang. This involved pointing the Hubble Space Telescope for several days at an empty part of the sky about one tenth as large as the full moon. The infrared Hubble Ultra Deep Field pushed to about 600 million years after the Big Bang and maybe fainter to redshifts as large as 10. This is the state of the art. Hubble's successor, the James Webb telescope which we shall discuss at length in Chap. 12 is expected to probe to redshifts of 20, less than 200 million years after the Big Bang (Fig. 9.1). Two hundred million years is a very long time in our everyday world but it is a small fraction (1.5 %) of the age of the universe. By exploring this new territory we also expect to discover entirely new phenomena. This has been the pattern throughout the history of astronomy.

The science of geology is similar to observational cosmology. Geologists study the sedimentary layers of the Earth's rock to tell the story of the Earth across hundreds of millions of years of its history. In the southwestern United States the rocks tell a remarkable story of river deltas and the ebb and flow of vast inland seas in stark contrast to the desert landscapes which we see around us today. In a similar manner our telescopes show us that galaxies in the distant past were quite different from the ones we see around us today.

Cosmology is based on the assumption that we live in an ordinary region of the universe and that, averaged over large enough volumes, the universe is essentially the same everywhere. If this is the case then observations of very faint distant galaxies seen at very early times give us a picture of what our own galaxy and its neighbors would have looked like at comparably early times. The challenge in studying these very distant galaxies lies in the fact that they are extremely faint and that their visible light is shifted into the infrared by the time it reaches our telescopes.

We can use galaxies to map the rise and fall of star formation over 95% of the history of the universe. Charting the cosmic history of star formation tells us how galaxies formed and how the universe evolved with time. As we push further back we can hope to see the epoch of reionization, the 600 million year era when the first stars and galaxies formed and ionized the neutral hydrogen in the universe.

In this chapter we present the methods used to detect the highest redshift galaxies, we will show examples of galaxies that have been discovered with Hubble Space Telescope observations. We then show how we can study the cosmic history of star formation. Our Milky Way galaxy forms stars at a rate of three solar masses per year, but the results indicate that galaxies formed stars at a much higher rate in the past. We discuss in the final section attempts to find out if the known number of distant galaxies is sufficient to ionize the universe or whether we have to probe even further back in time to discover the first galaxies to form after the Big Bang at the end of the dark ages.

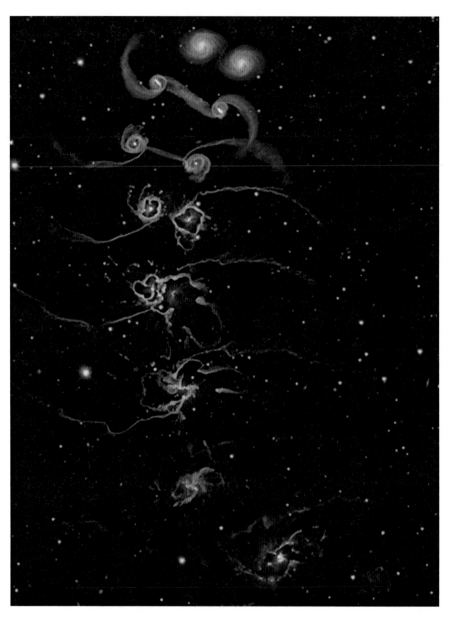

FIG. 9.2 Computer simulation shows two galaxies colliding. Color indicates temperature, and brightness indicates the gas density. When the central black holes merge, a quasar is ignited, pushing gas outward (Credit: Tiziana Di Matteo, Carnegie Mellon University)

When Did the First Galaxies Form?

Observations of massive elliptical galaxies in our neighborhood suggest they formed a long time ago since they are composed of old stars that have an age comparable to the age of the universe. In contrast, lower mass galaxies such as dwarf galaxies and spiral galaxies are still forming stars and contain old and young stars. However our theories of dark matter structure formation (Chap. 6) predict that the lower mass halos form first and the more massive ones later by mergers.

A second problem is that the ratio of gas to dark matter in galaxies is several times smaller than the ratio of gas to dark matter in the universe as a whole. This can only mean that gas must have been expelled from galaxies, but how? We see evidence for winds expelling gas from galaxies at redshifts of two and lower. We think these winds are produced by the radiation from young stars in star-forming regions of galaxies, although supernova explosions and black holes also contribute. Figure 9.2 illustrates how black holes might expel gas during a merger of two galaxies. These processes are known as feedback processes and are being modeled using computer simulations.

The current picture is that massive elliptical galaxies form in two stages. At high redshift gas enters dark matter halos along filaments known as cold streams and turns into stars at the center of the halo. Eventually the buildup of hot gas through feedback processes shields the halo from further gas inflows. These compact young galaxies are dominated by young blue stars. They eventually age and merge to produce more massive elliptical galaxies in the second stage of galaxy formation. To see if these ideas have any validity we must turn to observations.

Searching for the First Galaxies

We have opened a window into the first billion years of the universe to see the beginning of the growth and evolution of cosmic structure. The scientific motivation for this difficult work is to measure the cosmic history of star formation at the earliest times and to detect the objects that reionized the universe. But how are we to detect these very faint galaxies at the edge of time? We can't easily identify them in a single image of the sky because the vast majority of galaxies have redshifts much less than five. We need some way to sort the most distant objects from the nearby ones. Three methods are discussed below.

The first method makes use of the fact that hydrogen in high redshift galaxies and their surroundings absorbs almost all the short wavelength ultraviolet light. This method selects galaxies that contain mostly young stars and have little dust.

A second method makes use of two features in the spectra or colors of galaxies that are evidence of the presence of stars older than half a billion years. These features are known as the Balmer break and the 4,000 Å break. These features at redshifts of five and higher get redshifted to infrared wavelengths that are accessible from space. We must await the launch of the James Webb Telescope to arrive at reliable results for galaxies at redshifts of five and above containing older stars.

The third method uses far infrared and sub-mm emission from dust in high redshift galaxies. The very first galaxies and stars are not expected to contain dust since the chemical elements contained in dust are only produced inside stars. However we do not know at what stage dust makes its first appearance in the observable universe. This method is still in its infancy at high redshifts but new results are expected from a new telescope, the Atacama Large Millimeter/submillimeter array which has started carrying out observations high in the Andes.

Star Forming Galaxies at High Redshift: Lyman-Break Galaxies

The Lyman-break method selects high redshift galaxies that are forming lots of young stars. Some of the newborn stars are more massive than the Sun and emit large amounts of light at ultraviolet and blue wavelengths. As we discussed in Chap. 8, the high redshift universe contains enough neutral hydrogen to absorb ultraviolet light that may be emitted by these galaxies. The details of absorption of light by atomic hydrogen depend on the density of those atoms. At redshifts less than three, the hydrogen atoms

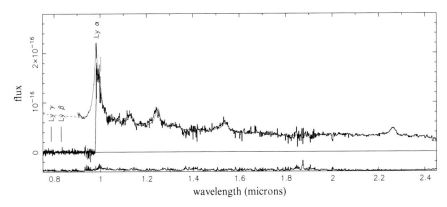

FIG. 9.3 The near-infrared spectrum (*black line*) of a distant quasar at a redshift of 7.1. The quasar is observed only 0.77 billion years after the Big Bang, but it has an intrinsic spectrum very similar to the averaged spectrum of lower redshift quasars (*red line*). The key difference is that shortward of a wavelength of about 1 μm corresponding to an emitted wavelength of $\lambda_{rest} \simeq 1,216$ Å there is no light. The larger amount of neutral hydrogen along the line-of-sight has absorbed all the ultraviolet light emitted by the quasar. This sudden drop in emission at wavelengths shorter than of Ly-α enables us to select rare extreme-redshift quasars such as this, but also more numerous "Lyman-break galaxies" at redshifts $z > 5$ (Credit: Reprinted by permission from Macmillan Publishers Ltd: Nature, 474, 7353, A luminous quasar at a redshift of z = 7.085, Daniel J. Mortlock et al. copyright (2011))

absorb light that has a wavelength shorter than 912 Å. As we go to redshifts greater than five, the absorption starts at wavelengths of 1,216 Å the wavelength of the Lyman-α line. We can use this effect to identify high redshift galaxies. This technique is illustrated in Figs. 9.3–9.5. Figure 9.3 shows the spectrum of a quasar at a redshift of 7.1. The light is emitted by fast moving hot gas surrounding a massive black hole. The point to note is that all the light emitted by the quasar at wavelengths shorter than 1,216 Å is missing because it has been absorbed by neutral hydrogen atoms. Bear in mind that the light emitted by the stars in the ultraviolet at 1,216 Å gets redshifted to a wavelength of 1 μm in the infrared by the time it reaches our telescope. The result is that the light detected at wavelengths shorter than 1 μm is missing in Fig. 9.3.

We can select galaxies in images using this effect as shown in Fig. 9.4. We measure the colors of galaxies by taking images of them with filters placed in front of the camera that let through only light of a specific range of wavelengths. The light profile or throughput of these filters is shown in the figure. For example, the

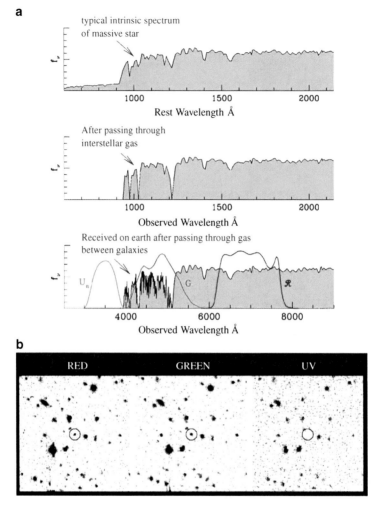

FIG. 9.4 The Lyman-break method for finding galaxies at redshifts of about three. The sharp break expected at wavelengths shorter than 912 Å in the spectrum of a galaxy dominated (in the UV spectrum) by massive, young stars is shown (a *top*). The break is accentuated by absorption both in the galaxy hosting the stars (a *middle*) and in the intervening intergalactic medium (a *bottom*). The galaxy redshift is 3.1 in this example, bringing the feature into the optical window observable from the ground. Shown in **a** Bottom are broad-filter passbands that can be used to find Lyman-break galaxies in the vicinity of redshift z = 3. (**b**) An example of images taken through these filters. The circled galaxy is seen clearly through the *red* and *green* filters, but it disappears completely through the UV filter. Only a few percent of all comparably faint galaxies will behave in this way (Credit: Charles C. Steidel, Observing the epoch of galaxy formation, Proc. Natl. Acad. Sci. USA, Vol 96, issue 8, April 13 1999, copyright (1999) National Academy of Sciences, U.S.A)

filter labeled R lets through light in the wavelength range 6,000–7,500 Å. For a redshift three galaxy the light with a wavelength less than 912 Å has been completely absorbed by neutral hydrogen, however if any light were present we would see it redshifted into the short wavelength visible band known as the U-band. But it is not present. The consequences of this are that when we take a series of images in the various filters, the image of the galaxy in the U-filter will not be present. Such galaxies are referred to as U-band dropouts. The U-dropouts are galaxies in the redshift range two to three. Galaxies at higher redshifts will be observed as dropouts in longer wavelength filters. For example B-dropouts are galaxies in the range three to four because the B-filter lets through light of longer wavelengths than the U-filter. One of the reasons for building the James Webb Space Telescope is to push the dropout technique into the infrared to find even more distant galaxies.

Why not simply obtain spectra of galaxy candidates and decide the redshift from the many emission and absorption features that are seen in the spectra? This is difficult because these galaxies are very faint. The galaxies are barely visible in images (Fig. 9.5) and obtaining a spectrum requires much more telescope time than obtaining an image. Figure 9.6 shows the thousands of galaxies detected in the Hubble Ultra Deep Field image. We cannot tell by inspecting this image which galaxies have the highest redshifts but by taking an image with different filters we can use the dropout technique to select the most distant galaxies ever seen with a telescope.

As we search for ever more distant galaxies, the dropout technique shifts to longer and longer wavelengths eventually moving out of the visible and into the infrared part of the spectrum. Using the infrared filters on the Hubble Space Telescope we can push the dropout technique out to redshifts of seven as illustrated in Fig. 9.5. This is one reason the study of the youngest and most distant galaxies depends on sensitive infrared detectors on large aperture telescopes on the ground or in space.

The dropout technique is not completely reliable. At a redshift of six, the technique produces one in four misidentifications; galaxies at lower redshifts that are mistaken for higher redshift galaxies.

There is also a population of galaxies known as Lyman-alpha blobs that are the opposite of the dropouts, these galaxies actually emit Lyman alpha radiation. These galaxies do have in common with the dropouts that they are star-forming galaxies. Why are not all star forming galaxies Lyman-alpha emitters? It appears that this depends on the details of the structure of these galaxies.

Star-Forming Galaxies at High Redshift: Sub-millimeter Galaxies

Emission in the far infrared that is not associated with the cosmic background radiation comes from galaxies that contain dust. The dust re-emits light in the infrared that it has absorbed at optical wavelengths. In fact one can show that for every two photons of light emitted by stars or by black holes one photon has been absorbed by dust and re-radiated as infrared and sub-millimeter emission by the dust. Since dust is closely associated with star-forming regions in galaxies, we can use dust emission as a tracer of star formation.

The dust emission spectrum has an interesting property that observers of distant galaxies can exploit. The intensity falls steeply from a peak of wavelength of one tenth of a millimeter down to about 2 mm. The emitted energy drops by a factor of about 1,000. Consequently sub-millimeter galaxies can appear brighter in our telescopes when observed at increasingly high redshifts. This is because for a fixed observing wavelength the emitted wavelength decreases with increasing redshift. These simple calculations show that the millimeter wavelength range is a good place to search for very distant galaxies that are actively forming stars.

One of the first cameras to operate at these wavelengths was the SCUBA (Sub-millimeter Common User Array) on the James Clerk Maxwell Telescope at Mauna Kea observatories in Hawaii. The first studies made with this camera revealed the existence sub-millimeter galaxies. The location of sub-millimeter galaxies on the sky could not be determined to high enough accuracy to see if optical counterparts to these galaxies exist in the Hubble Deep Fields. By combining SCUBA observations with infrared and radio wavelength observations it has become possible to identify these galaxies optically and take their spectra at visible wavelengths using the Keck telescopes.

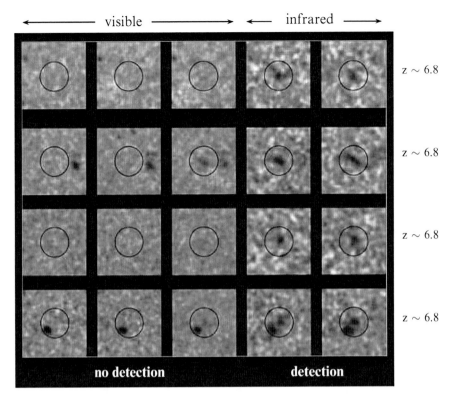

visible ⟶   ⟵ infrared ⟶

z ∼ 6.8

z ∼ 6.8

z ∼ 6.8

z ∼ 6.8

no detection   detection

FIG. 9.5 Four candidate galaxies that are likely to have redshifts of 7 and thus have emitted their light when the universe was just 750 million years old. Each of the four candidate high-redshift galaxies are presented in a distinct row. All four candidate galaxies are shown using images at each of five different wavelengths. The three columns starting from the *left* are images taken with filters that select colors in visible light with bluer light on the *left* and redder on the *right*. The last two columns are images taken with two infrared filters. The galaxies are all detected in the infrared but remain completely undetected at wavelengths shorter than red visible light. This abrupt drop-off in light emission is characteristic of star-forming galaxies at high redshifts and occurs due to the absorption of light by the large amounts of neutral hydrogen in the universe at early times. Astronomers use the presence of this break to find high-redshift galaxies. The present sources were found in the faintest images taken with the Hubble Space Telescope (Credit: Rychard Bouwens)

Keck telescope data revealed that the sub-millimeter galaxies are at redshift two on average (i.e. the light was emitted at a time when the universe was about 3 billion years old). These galaxies occurred more frequently in the past and make a significant contribution to the star formation at high redshift.

FIG. 9.6 The Hubble Ultra Deep Field. There are about 10,000 galaxies in this image. Using the Lyman break or dropout techniques described in the text we can search for the most distant galaxies out to redshifts as high as ten using these data (Credit: NASA/STScI)

It is possible that the sub-millimeter galaxies are merging galaxies that contain alot of gas. This gas is then used up in a huge burst of star formation. The sub-millimeter galaxies and the Lyman-break galaxies have huge rates of star formation from several hundred to several thousand solar masses of stars formed per year. Our Milky Way galaxy by comparison forms stars at a rate of a few solar masses per year.

Older Galaxies at High Redshift: The Distant Red Galaxies

We discuss below the tools for discovering old galaxies at high redshifts and the implications of these discoveries. By comparing observed galaxy spectra to models constructed from stellar spectra one can learn something about the ages of galaxies. A galaxy spectrum consists of the light from lots of stars of different mass and age, each with its own spectrum. If we imagine a bunch of stars all forming at the same time (the simplest assumption) we can predict how the spectrum of such a galaxy will change as it gets older. We know that the more massive hot blue stars run out of hydrogen soonest and explode as supernovae. As the population ages and the blue stars contribute less light, the galaxy spectrum changes. Figure 9.7 shows how the spectrum of a galaxy is expected to change as its population of stars age.

There is a feature in the blue part of the spectrum of galaxies that we can use to detect galaxies with older stars. This feature is called the 4,000 Å break. 4,000 Å refers to wavelength of the blue light where the feature occurs. The feature is caused in part by the absorption of light by ionized calcium. The feature is also more pronounced because the light from old stars is dropping off towards the blue.

Galaxies with an old population of stars have been identified at high redshift using the 4,000 Å break. At a redshift of three this feature is redshifted into the infrared part of the spectrum (1.6 μm). We can select older galaxies in the redshift range $2 < z < 3$ by imaging them at 1.24 and 2.2 μm using standard infrared filters. Spectra of galaxies selected in this manner do not show emission lines and appear to have an old population of stars.

Figure 9.7 illustrates how the spectrum of a galaxy can change with time due to the effects of an aging stellar population. Figure 9.8 shows a model of a redshift seven galaxy spectrum. The figure shows the colors of the light that a redshift seven galaxy might have when observed by the Spitzer and Hubble space telescopes.

The distant red galaxies described above may teach us about the origin of the elliptical galaxies we see around us today. Galaxies discovered in the redshift range $1.4 < z < 2$ formed most of their

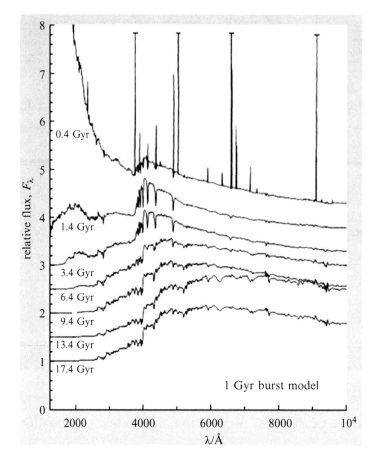

FIG. 9.7 Simulated galaxy spectra following a billion year (Gyr) burst of star formation. Ages are marked in Gyr on each spectrum. Note the dramatic change in the relative amounts of flux blueward and shortward of 4,000 Å as the galaxy ages. The galaxy will appear to get progressively redder as it ages (Credit: Stephen Serjeant, Observational Cosmology, Cambridge University Press, which has been adapted from Fig. 1 in Rocca-Volmerange & Guiderdoni, 1988, A&AS, 75, 93, reproduced with permission ©ESO)

stars about 1 billion years earlier at redshifts of about three. These objects then merge to form the massive elliptical galaxies that we see today.

What Do High Redshift Galaxies Look Like?

Figure 9.9 shows a set of images of distant galaxies seen in the Hubble Ultra Deep Field. These objects look quite different to any

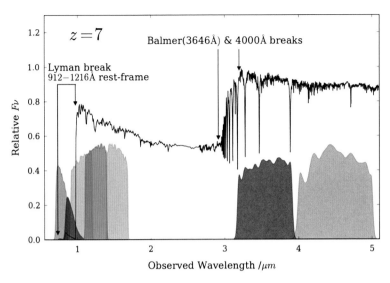

FIG. 9.8 The redshifted light anticipated from a young galaxy at redshift seven. The emitted ultraviolet light can be sampled by the red color and near-infrared filters on the Hubble Space Telescope shown in *blue* and *green* in the Figure. The longer wavelengths can be probed with filters aboard an infrared satellite called Spitzer. The Spitzer filters are shown in *red* in the figure. The galaxy light shows the sharp drop at $\lambda_{rest} = 1,216$ Å due to the strong absorption by intervening neutral hydrogen anticipated at this redshift. Longward of this "Lyman-break" the galaxy light shown is simply that of all the stars added together. The features known as the Balmer break and the 4,000 Å break can help estimate the age of the stars in the galaxy. As the galaxy ages these feature become more pronounced as shown in Fig. 9.7. The gap between the Hubble Space Telescope and the Spitzer Space Telescope filter coverage will not be covered from space until the advent of the James Webb Space Telescope (Credit: A. B. Rogers and J. S. Dunlop)

galaxy seen nearby. The first impression is that the high redshift galaxies are more lumpy than nearby galaxies. Scientists must always examine initial impressions critically. Winnie-the-Pooh, who wasn't a scientist, thought that he had discovered in the snow the footprints of a Heffalump when the footprints were really his own. These Heffalump detections also occur in astronomy.

When we use a filter of given wavelength to image a galaxy, the wavelength at which the light that we observe was emitted gets shorter and shorter for high redshifts. We must be careful to compare images of high and low redshift galaxies at similar emitted wavelengths. At high enough redshifts an infrared image of a distant galaxy must be compared with an ultraviolet image

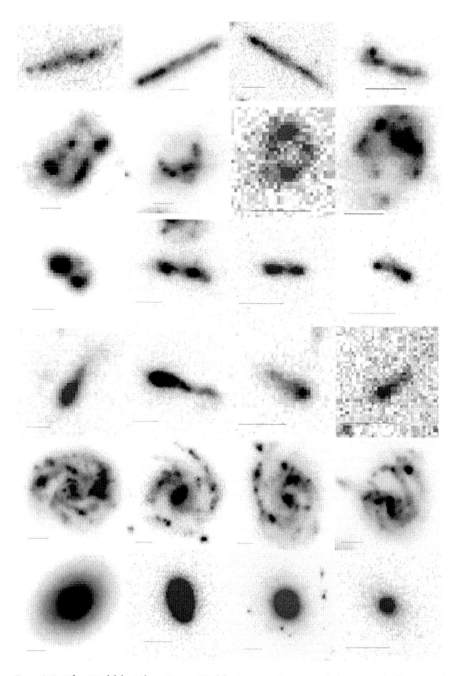

Fig. 9.9 The Hubble Ultra Deep Field gives a glimpse of the morphologies of galaxies at high redshifts. These morphologies are clumpy and irregular (*top three rows*) in many cases in contrast to spiral and elliptical galaxies. We see *top row* to *bottom row*: chain morphologies, clump-cluster, double, tadpole, spiral and finally elliptical galaxies (Credit: Debra Elmegreen et al. 2005, Astrophysical Journal, 631, 85, Reproduced by permission of the American Astronomical Society)

of a nearby galaxy. Secondly, the distant galaxy images will only reveal the brightest features that we would see in nearby galaxies. Even when these observing biases are taken into account the fact remains that high redshift galaxies look very different to their low redshift counterparts.

High redshift galaxies have been visually classified into categories such as tadpole, double, chain, chain cluster. None of these morphologies are seen in low redshift galaxies. The clumps seen in high redshift galaxies have masses of 100 million solar masses in stars and sizes of 3,000 light years or less.

We are not quite sure what to make of these unusual galaxy shapes. It is possible that we are witnessing the formation of galactic disks by mergers of smaller halos. Maybe we are witnessing the rapid cosmological infall of gas into a halo. The images suggest that the clumps are regions of intense star formation in disk galaxies.

Cosmic Accounting: Star Formation and the History of the Universe

Can the number of stars we see locally in our universe can be accounted for by counting the number of stars formed since the Big Bang? The lifetimes of stars are much longer than those of human beings. A star having the mass of our sun is expected to shine at roughly constant brightness for about 10 billion years. We thus have a consistency check. We expect the birth rate of stars, that is the star formation rate added up until a certain time to be equal to the density of stars seen at that time.

How then are we to measure the star formation rate of the universe? We measure this as a density since we can't observe the entire volume of the universe. We take a very large volume of space and use our astronomical observations to estimate how many stars are forming in that volume per year. We then do the same exercise at various redshifts going further back in time. There are various methods used to estimate star formation in galaxies:

The first method is to measure the ultraviolet light emitted by galaxies. Most of that light is emitted by stars having more than five times the mass of the sun. The stars that emit most of the ultraviolet flux only live a few million years so they can't have been around very long. Their mere presence is an indicator that stars were forming in the 'recent' past. We do have to make assumptions about the removal of ultraviolet light by dust to derive the star formation rate from these measurements.

Secondly, we know that recently formed massive stars are hot enough to ionize gas clouds which will emit light. We see the resulting emission lines in the spectra of galaxies, particularly the elements oxygen and hydrogen.

Thirdly, far infrared emission is caused by dust heated by the light of young stars; we can use this infrared flux as a measure of the star formation rate.

The fourth method uses the radio wavelength emission from galaxies to estimate the amount of recent star formation. A 10 solar mass star has a lifetime of 10 million years, one 10,000th of the lifetime of the Sun. These massive stars end their lives in tremendous explosions known as supernovae. What remains after such an explosion is an expanding gas cloud known as a supernova remnant. These supernova remnants are known sources of radio emission, there are some spectacular ones in our own galaxy. In practice we can use the brightness of a galaxy at 20 cm wavelengths to estimate the star formation rate from the supernova rate.

If these different measurements are all caused by the same phenomenon (star formation) they should be correlated with each other; galaxies that are brighter at 20 cm radio wavelengths should also be brighter in the far infra-red. This is indeed the case.

We can use these measures of star formation rate to estimate star formation history from redshift zero (today) all the way back to high redshifts (Fig. 9.10). In doing this we are plotting the star formation rate over 90 % of the age of the universe. It turns out the star formation rate was much higher 10 billion years ago than it is today. We can also measure the number of stars per unit volume (the stellar mass density) out to redshift four. The stellar density is measured by the near infrared luminosity of these galaxies. It turns out that the stellar mass density increases by a factor 20 from redshift 4 down to redshift 0.

Is this increase in the number of stars in our universe consistent with the rate at which stars have been forming over the past 12 billion years? The answer appears to be yes at low redshifts. The addition of all the stars formed in the past 12 billion years accounts rather well for the density of stars we see in the universe

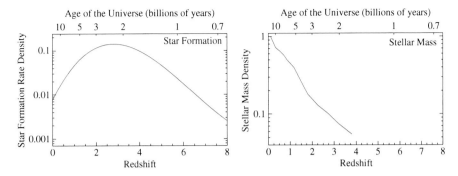

FIG. 9.10 *Left panel*: the star formation density history of the universe. These measurements are derived from mid-infrared luminosities and sub-mm luminosities of galaxies. *Right panel*: the stellar mass density of the universe as a function of time and redshift. The star formation rate observed in the *left panel* is greater by a factor five from what one would infer from the stellar density measurements

today. Half of the mass of stars we see today had already formed by redshift of two, when the universe was 3 billion years old. This is consistent with measurements of star ages in our galaxy. The typical stars in our Milky Way disk were formed 3–6 billion years ago, whereas the oldest stars in our galaxy were formed at least 10 billion years ago.

The density of stars today is computed by adding up the near-infrared luminosity of all the galaxies we see in a nearby volume of space. We choose to use the near-infrared as a tracer of mass in stars. Infrared light is not subject to obscuration by dust. We have to correct for the fact that some of the mass that was formed into stars has returned to the space between the stars. There are two main causes for this, supernova explosions and stellar winds which can cause stars to loose their outer layers. These effects are significant; 30% the mass can be lost in this way. The infrared measurements reveal that about 7% of the baryons in the universe are in the form of stars and the rest are mostly in the form of ionized gas.

At redshifts larger than five it becomes more difficult to estimate the star formation rate and the density of stars and conclusions about consistency and the effects of dust absorption are less reliable.

Cosmic Accounting: The End of Reionization

The reionization of the hydrogen gas in the universe marks the end of the cosmic dark ages and the beginning of the age of galaxies. This is one of the key transitions in the history of the universe yet we know very little about this period. We do know from the spectra of quasars that the universe was reionized by redshift 6. Observations of the cosmic background radiation suggest the process started at redshift 15 when the universe about 300 million years old.

Where are the sources that reionized the universe? To ionize a hydrogen atom takes an ultraviolet photon. After it is ionized the hydrogen atom will recombine with the electron to form a neutral atom. For a gas of known density and temperature we can calculate the rate at which this will happen. To keep the universe ionized the production of ultraviolet photons capable of ionizing atoms has to match or exceed the rate at which protons and electrons can reform into atoms.

At redshifts less than six the ultraviolet radiation from quasars is sufficient to keep the universe ionized. However at redshifts greater than six, there are not enough quasars around to provide the amount of ultraviolet light needed to ionize the universe. What then is keeping the universe ionized at redshifts higher than six?

Recent observations of galaxies at redshifts of seven suggest that galaxies rather than quasars may have provided the ultraviolet radiation that kept the universe ionized. This then is our second accounting problem. Can we detect the objects responsible for ionizing the universe?

For a star of given mass we know how much ultraviolet light is produced during the lifetime of the star. One can then estimate how many of these stars are required to ionize a given volume of space. Since stars are located within galaxies, we also need to know what fraction of the ultraviolet light can escape the galaxy and ionize the hydrogen atoms between the galaxies.

By measuring the ultraviolet flux from galaxies we can estimate the rate of star formation up to redshifts as high as seven

(Fig. 9.10). We also know from Hubble Space Telescope infrared observations how luminous these galaxies are. It seems that most of the ultraviolet light at high redshifts is produced by a large number of very faint galaxies. To find out whether these galaxies produce enough light to keep the hydrogen in the universe ionized new observations of faint distant galaxies are required. We need to image galaxies that are two to three times fainter than the galaxies seen in the longest exposure taken with the Hubble Space Telescope.

The ultraviolet light emitted by electrons going from level 2 to level 1 in hydrogen atoms (known as Lyman-alpha) may yield clues to the reionization question. Observations of high redshift Lyman-α emission from galaxies may help establish when reionization began. At higher redshifts, the fraction of galaxies with detectable Lyman-α emission increases. This suggests that at higher redshifts galaxies have lower dust content as one would expect of the first galaxies. When we go far enough back in time that the universe turns neutral, the Lyman-α emission should be absorbed and this would document the beginning of reionization.

Review

We have presented several techniques that are used to find the most distant galaxies known to us. We have not yet found the first galaxies but we are getting close. We can use the known high redshift galaxies to estimate how the star formation rate changes with redshift. We can also estimate the density of stars at various redshifts. It is instructive to do a consistency check using these quantities, since the sum of the star formation rate over the history of the universe can be used to independently compute the density of stars at any given time. The results seem consistent when dust absorption is taken into account. We have not yet detected the sources of ultraviolet light that are keeping the universe ionized above a redshift of six. A number of puzzles remain, but we are beginning to understand how galaxies were assembled over the history of the universe.

Further Reading

High-Redshift Galaxies. I. Appenzeller. Springer, 2009.

From First Light to Reionization. M. Stiavelli, Wiley-VCH, 2009.

First Light in the Universe. A. Loeb, A Ferrara and R.S. Ellis, Springer, 2008

Observational Cosmology. Stephen Serjeant, Cambridge, Cambridge University Press, 2010

Observing the First Galaxies. James Dunlop, in The First Galaxies - Theoretical Predictions and Observational Clues, Springer, eds. V. Bromm, B. Mobasher, T. Wiklind 2012

10. Cosmic Archaeology

At first I could see nothing, the hot air escaping from the chamber causing the candle flame to flicker, but presently, as my eyes grew accustomed to the light, details of the room within emerged slowly from the mist, strange animals, statues, and gold - everywhere the glint of gold. For the moment - an eternity it must have seemed to the others standing by - I was struck dumb with amazement, and when Lord Carnarvon, unable to stand the suspense any longer, inquired anxiously, "Can you see anything!" it was all I could do to get out the words, "Yes, wonderful things".

Howard Carter, The Tomb of Tutankhamen

The ancient light that we gather with the largest ground based telescopes together with Hubble Space Telescope data reveals the appearance of galaxies soon after their formation. These galaxies are so distant that we cannot see them in as much detail as nearby galaxies. We can also learn about the distant past by studying nearby galaxies. We can study the age and chemical composition of their stars and use these to reconstruct the history of these systems. In this sense, nearby galaxies are like fossils. In this chapter we present evidence about galaxy formation that we obtain from the study of galaxies very close to us. By close, we mean distances of a few million light years as opposed to the billions of light years that separate us from the galaxies discussed in the previous chapter.

The closest galaxies to the Milky Way lie in a structure known as the Local Group. The local group contains three large spiral galaxies surrounded by numerous dwarf and irregular galaxies. These spiral galaxies are the Milky Way the Andromeda Galaxy and the Triangulum galaxy. We will show evidence of galaxies merging with our Milky Way galaxy and getting ripped apart by tidal forces in the process. We will also show how the populations of stars in nearby dwarf galaxies can be used to measure the

G. Rhee, *Cosmic Dawn: The Search for the First Stars and Galaxies*, Astronomers' Universe, DOI 10.1007/978-1-4614-7813-3_10, © Springer Science+Business Media, LLC 2013

formation history of those galaxies. The number of small galaxies surrounding the Milky Way must be explained by our theories. Dwarf galaxies are of interest to us because they represent the extreme faint end of galaxy formation.

The Milky Way Galaxy

The Milky Way has been known for many centuries as a feature of the night sky. The ancient Egyptian and Greeks were aware of it as were native tribes in the Andean foothills and rain forests of south America. Over 400 years ago Galileo used his telescope to reveal that the Milky Way consists of huge numbers of stars. In the late eighteenth century, William Herschel and his sister Carolyn attempted to map the Milky Way by counting stars in different parts of the sky, revealing the Milky Way to be a flattened structure. In 1917 Harlow Shapley calculated distances to a 100 globular star clusters and used these to locate the center of the Milky Way. He concluded that our solar system is located about two thirds of the way from the center of the Milky Way disk. New observing techniques developed in the twentieth century made it possible to observe the Milky Way across the electromagnetic spectrum. At near infrared wavelengths we can see through the dust and see into the disk of the galaxy (Fig. 10.1). Far infrared wavelengths reveal the location of the dust. Radio observations give a detailed view of the neutral hydrogen gas that is distributed in and above the plane of the galaxy.

The Milky Way is a spiral galaxy with four major components. The first is the dark matter halo whose presence is revealed through its gravity. The total mass of the galaxy is a 1,000 billion (10^{12}) times the mass of the Sun. The second major component is the disk which is about 100,000 light years in diameter and about 3,000 light years thick. The whole disk is embedded in a halo of stars. The disk alone contains 100 billion (10^{11}) stars. It takes 200 million years for our solar system to orbit the center of the galaxy. The fourth component is a bulge in the center of our galaxy that is part of a bar extending from the center.

The cosmic background measurements tell us that the ratio of baryonic (atomic) matter to dark matter in the universe is 1/5. The observed ratio, mass in stars to total mass, for our Milky

FIG. 10.1 The entire sky as seen by Two Micron All-Sky Survey. The measured brightnesses of half a billion stars (points) have been combined into colors representing three distinct wavelengths of infrared light: *blue* at 1.2 μm, *green* at 1.6 μm, and *red* at 2.2 μm. This image is centered on the core of our own Milky Way galaxy, toward the constellation of Sagittarius. The two faint smudges seen in the lower right quadrant are our neighboring galaxies, the Small and Large Magellanic Clouds (Credit: Infrared Processing and Analysis Center, Caltech and University of Massachusetts)

Way galaxy is 1/20. In other words we see far fewer stars than we would expect based on cosmic measurements. This effect is more pronounced for dwarf galaxies which are even less efficient at forming stars out of their gas than Milky Way sized galaxies.

This cosmic accounting relates the basic components of the mass in the universe (dark matter and atoms) to the observed properties of galaxies (dark matter halos, stellar disks, gas and dust). Our theories must explain how the dark matter clumped into structures, how the gas condensed in these dark matter clumps and how stars were formed. The star formation efficiency or fraction of gas that turned into stars is a number that depends on the mass of the dark matter halo within which a galaxy finds itself. Current theories cannot account for this number yet.

The Milky Way according to the cold dark matter theory of structure formation should be surrounded by many hundreds of small galaxies but we only see about 50 in our observations. A challenge is to reconcile the large predicted number of low mass concentrations with the smaller number of observed dwarf galaxies. The theory also predicts that merging of smaller galaxies with

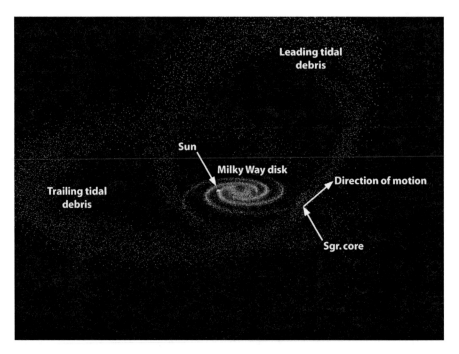

FIG. 10.2 The distribution of stars in the shredded Sagittarius dwarf galaxy as revealed by observations. The image is based on the best model match to the map of 2MASS M-giant stars. The thin flat *blue spiral* represents the disk of our Milky Way galaxy. The *yellow dot* represents the position of the Sun. Sagittarius debris can be seen extending from the dense 'core' of the Sagittarius dwarf, wrapping around the galaxy, and descending through the Sun's position (Credit: David Law/UCLA)

the Milky Way should be taking place today. We do see evidence in the Milky Way halo of past mergers with dwarf galaxies. The visible consequences of such a merger are illustrated in Fig. 10.2.

The Andromeda Galaxy

The most distant object that can be seen with the naked eye is the Andromeda Nebula also known as M31. It is the nearest large galaxy to our own at a distance of 2.5 million light years. The disk of the galaxy spans several moon diameters on the sky. The disk is round like a dinner plate but appears elliptical on the sky because it is not viewed directly face on but rather at an angle to our sight line.

FIG. 10.3 Viewed in ultraviolet light, M31 looks more like a ring galaxy than a spiral. The ring is highlighted in this image taken by the GALaxy Evolution Explorer (GALEX) satellite. Ultraviolet colors have been digitally shifted to the visual. Young blue stars dominate the image, indicating the star forming ring as well as other star forming regions. The origin of the huge 150,000-light year ring is unknown but likely related to gravitational interactions with small satellite galaxies that orbit near the galactic giant (Credit: GALEX team, Caltech, NASA)

Like our own galaxy, M31 has been imaged from the ultraviolet to the radio part of the spectrum. Ultraviolet images (Fig. 10.3) reveal the presence of young hot stars. The near infrared enables us to map the old star population, the far infrared emission is a tracer of dust and the radio emission lets us map the neutral hydrogen gas. In the infrared, the dust reveals the presence of a ring of star formation with spiral structures extending inwards. The ring is also seen in the ultraviolet. Simulations suggest that the ring formed when a companion dwarf galaxy plunged through the center of M31 about 200 million years ago. In the simulations, the encounter destroys the spiral arms of M31 and brings about the formation of a ring-like structure. M31 is moving towards the Milky Way and the two galaxies will collide in about 3 billion years. Simulations of this event indicate a large probability that the Sun will be kicked out to distances a few times larger than its current distance from the Milky Way. In a few billion years the night sky seen from earth may look very different than it does today with the Milky Way being visible to anyone on earth as an external galaxy.

The Local Group

Swarms of dwarf galaxies surround the Milky Way and M31. They form a structure about 6 million light years in diameter known as the local group. It is a real structure in the sense that its member galaxies are participating in the expansion of the universe. The local group is shown schematically in Fig. 10.4. The local group is what we call a bound structure, held together by its own gravity. The next most massive galaxies after the Milky Way and M31 are the spiral galaxy M33 and the small and large Magellanic clouds. Most of the remaining galaxies are dwarf galaxies with masses of just a few percent of the Milky Way.

The faintest known dwarf galaxies have luminosities ten million times smaller than that of the Milky Way. These galaxies are located in the lowest mass dark matter clumps that are known to host stars. These galaxies are so small that we can ask whether they should not be referred to as star clusters. However, if we define galaxies as consisting mostly of dark matter, then these objects are dwarf galaxies. These faint objects are very difficult to detect against the bright Milky Way. Large digital surveys of the sky in several colors have made this possible. Techniques have been developed (see Fig. 10.7) that have detected 14 new dwarf galaxies in the local group, taking the census up to 50. Local group studies have revealed facts of great interest to cosmology. These findings include the number of dwarf galaxies, the stellar population (color and brightness) in local group dwarf galaxies, and the structure of the halos of the two large galaxies M31 and the Milky Way.

FIG. 10.4 The location of the Milky Way galaxy in the local group. The two dominant galaxies are surrounded by many smaller galaxies a few of which are shown in this figure (Credit: NASA/CXC/SAO)

The Missing Satellite Problem

The currently favored model of cosmology is very successful at accounting for observations on the large scales of tens of millions of light years and larger. The model has encountered a number of problems on smaller scales. One such problem concerns the number of satellite galaxies of the Milky Way and Andromeda.

We use the motions of stars in dwarf galaxies to estimate the masses of these galaxies and thus the mass of their dark matter halos. We can then compare the observed number of halos of given mass with the model predictions. For the most massive dwarf galaxies the agreement is fairly good. Galaxies in the models with the mass of the Milky Way should have one or two companions as massive as the Magellanic Clouds. As we go to less massive halos a discrepancy emerges. There are many more halos in the dark matter simulations than there are dwarf galaxy satellites surrounding the Milky Way. In fact, our local group is expected to have about 1,000 small mass concentrations but we only observe 50 or so galaxies. The challenge is thus to reconcile the large number of low mass dark matter concentrations with the smaller number of observed dwarf galaxies. How then are we to solve this so called "missing satellite problem"?

We could modify the dark matter distribution so that gravity does not bring about the formation of low mass halos. This is not an option since independent observations of gas clouds constrain the dark matter distribution. If we accept that the predicted number of dark matter halos is correct, we can argue that not all halos form stars. Supernova explosions might prevent low mass halos from being visible to us. The first generation of stars are believed to be massive, they eventually explode as supernovas and the energy released in these explosions is sufficient to blow the gas out of low mass halos and prevent further star formation. The details of star formation and gas cooling suggest that most of the dark matter halos with mass less than 1% of the Milky Way halo should be completely dark. In fact, we may then encounter the opposite problem of explaining how the faintest detected dwarf galaxies could even exist. To solve this problem we need to know how efficient halos of given mass are at turning their gas into stars.

Star Formation in Dwarf Galaxies

Figure 10.5 shows the dwarf irregular galaxy NGC 1705. Some dwarf galaxies may be the first galaxies to have formed after the Big Bang. If the bigger galaxies formed by mergers of smaller galaxies, then the dwarf galaxies we see today may resemble the building blocks that assembled to form larger galaxies like our

FIG. 10.5 The dwarf irregular NGC 1705 is a small galaxy lacking regular structure. Young, blue, hot stars are strongly concentrated toward the galaxy's center. Older, red, cooler stars are more spread out. This galaxy has been forming new stars throughout its lifetime, but a burst of star-formation activity occurred as recently as 26–31 million years ago (Credit: NASA, ESA and the Hubble Heritage Team STScI/AURA)

Milky Way. With the technological leap provided by the Hubble Space Telescope it became possible to image individual stars in several local group dwarf galaxies. This made it possible to measure the history of star formation in these nearby dwarf galaxies by looking at their individual stars. This is done by plotting each star as a point on a plot of star brightness versus star color.

We can plot star color versus star brightness which allows us to estimate the relative numbers of older and younger stars.

Using Hubble Space Telescope data, star formation histories have been constructed for these galaxies and the results are quite

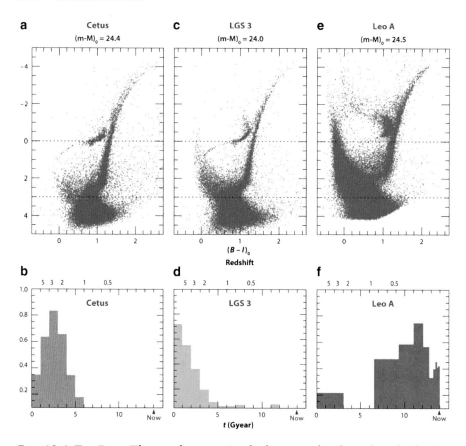

FIG. 10.6 *Top Row*: Three color magnitude diagrams for three dwarf galaxies. The horizontal axis plots the color, with *blue* on the left and redder towards the right. The vertical axis is magnitude with the brightness increasing along the vertical axis. Each point in these diagrams is a star, observed with the Hubble Space Telescope Advanced Camera for Surveys. On the *bottom row* we see the star formation rate (derived from the upper figure) plotted against time in billions of years (Gyr) with redshift labeled on the upper horizontal axis. Each column corresponds to a different galaxy's star formation history and color-magnitude diagram. As can be seen, the star formation histories can differ substantially from one galaxy to the next (Credit: E. Tolstoy, V. Hill and M. Tosi, Annual Review of Astronomy and Astrophysics, Volume 47, page 387, 2009, Star-Formation Histories, Abundances, and Kinematics of Dwarf Galaxies in the Local Group)

surprising (Fig. 10.6). Each column in the figure corresponds to a galaxy. The top row shows the star observations, and the bottom row shows the rate at which the galaxies formed stars over the history of the universe. Some galaxies like the Carina dwarf spheroidal galaxy show a number of distinct episodes of star formation with the star formation going to zero in between. Other galaxies such as Cetus show a long period of star formation lasting about 5 billion years after the Big Bang. In the dwarf galaxy Leo A, star formation happened over the last 8 billion years. It is unlikely that these dwarf galaxies in the local group are the building blocks of the Milky Way. True fossils are expected to have a single old stellar population or at least a temporary suppression of star formation after reionization. We also see that the element abundances in the stars in the dwarf galaxies do not resemble the abundances seen in stars in the Milky Way halo. The Milky Way and M31 could certainly have formed by merging of dwarf galaxies in the early universe, but the dwarf galaxies we see around today are quite different to those early building blocks. What then did the first galaxies look like?

Are the First Galaxy Fossils Lurking in the Local Group?

After the first atoms form in the universe the gas starts to fall into the dark matter halos. The gas reacts by increasing its temperature and pressure but keeps falling inwards if the time to build pressure is too slow compared to the time to fall into the clump. Stars from 20 to 100 solar masses could form in halos having a mass of one million solar masses. These massive stars produce enough ultraviolet radiation to ionize the gas in the halo and prevents further star formation. Eventually the first massive star explodes as a supernova and the gas can cool thanks to the presence of molecular hydrogen and the presence of heavy elements created by the first stars. This early period of star formation reionized the universe making it harder for low mass halos to form stars.

The first galaxies to form are believed to have masses between 100 million (10^8) and 1 billion (10^9) solar masses. These first galaxies will become dwarf galaxies. Between the formation of the first stars and the end of reionization it is possible that some galaxies form in halos less massive than 100 million solar masses. If these halos survive they will be galaxy relics from the pre-reionization era.

Could we observe these objects? The oldest known galaxies would be expected to be assemblages of a few hundred stars moving at speeds of about $20\,km\,s^{-1}$. These galaxy fossils as they are called could in fact resemble the least massive dwarf spheroidal galaxies that we see around the Milky Way. These faint ghostly objects are very hard to detect. The galaxy Leo T contains stars moving at about $7\,km\,s^{-1}$ in a dark matter halo of mass ten million (10^7) solar masses. There are several recently discovered candidate fossil galaxies. They are so faint that they can only be detected within 150,000 light-years of the Milky Way. More work is needed to confirm that these newly discovered objects are indeed fossils from the early stages of galaxy formation.

We have discussed in the previous chapter the search for the first galaxies at high redshift and in this chapter the search for nearby fossils of galaxy formation. The two methods produce quite different results. The distant searches reveal intrinsically bright massive objects that are forming stars at a very high rate. The redshift eight galaxies have stellar masses of one billion (10^9) solar masses in contrast with the ultra faint dwarf stellar masses of a few hundred solar masses. The nearby fossils may be in fact the building blocks of the bright massive galaxies we see in the distant universe, which themselves may be the ancestors of galaxies such as the Milky Way.

The Search for the Faint Galaxy Fossils

Large sky coverage digital surveys can be used to conduct computer searches for faint assemblages of stars that may be invisible to the eye. Prior to 1994, 10 galaxies were classified as satellites of the Milky Way, since then another 14 have been discovered. Based on these discoveries it is estimated that the actual total number of Milky Way satellites lies between 200 and 500. Note the large uncertainty!

The trick to finding these ultra faint companions of the Milky Way lies in finding an excess or overdensity of stars in a part of the sky. This is very difficult because the dwarf galaxies lie behind the Milky Way, and in any image of the sky the foreground Milky Way stars will greatly outnumber the stars in the dwarf galaxy. One has to apply a filter to eliminate foreground Milky Way stars and

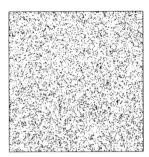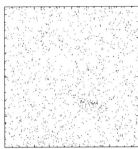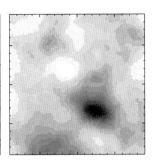

FIG. 10.7 *Left Panel*: Map of all the stars in the field around the Ursa Major I dwarf satellite. The field of view is about two moon diameters on the sky. The *center panel* shows the same field of view after removal of all stars except those that could plausibly be associated with a dwarf galaxy based on their color and brightness. The *right panel* shows a smoothed map of the star distribution in the center panel. The faintly visible dwarf galaxy is now clearly visible as the dark blob. The data are from the Sloan Digital Sky Survey (Credit: Beth Wilman, In Pursuit of the Least Luminous Galaxies, Advances in Astronomy, Volume 2010, page 4, 2010)

examine the remaining handful of stars. We make a guess as to what the colors and brightnesses of the dwarf galaxy stars would be and eliminate stars outside this range from the image. If the remaining stars are concentrated into a clump then one has found a candidate dwarf galaxy (Fig. 10.7). One then obtains spectra of the candidate stars to confirm whether they are members of a separate galaxy. The ultra faint galaxies found in this manner are as faint as star clusters in the Milky Way but larger in size. As we go fainter, dwarf galaxies become increasingly dark matter dominated. The spectra of the stars in these galaxies enable us to estimated abundances of elements such as iron relative to hydrogen. These dwarf galaxies have iron abundances much lower than that of our sun. Since iron can only be created inside stars, the iron abundance measures to what extent the gas in a star was processed in a previous generation of stars. This lack of iron suggests we are looking at very old stars that formed from gas that was not "polluted" by the gas ejected during supernova explosions.

The Large Scale Synoptic Telescope survey will make it possible to detect dwarf galaxies about eight times further than is currently possible. We should be able to find satellite galaxies right at the edge of the Milky Way dark matter halo. New radio telescopes such as the Square Kilometer Array telescope will search for dark matter halos that may contain only hydrogen gas.

Tidal Streams

The final steps on our archaeological tour of the local group are the halos of the two dominant spiral galaxies, our own Milky Way and M31, the Andromeda galaxy. Galaxies grow by accreting smaller galaxies, the small galaxies get disrupted as they approach the larger galaxy halo. The details depend on the halo mass and the speed and angle with which the victim approaches the large halo. We expect small satellites to be ripped apart by tidal forces as they approach the larger halo because the force of gravity due to the large halo is not the same at each location in the small halo. Different parts of the small halo experience different forces with the result that the small halo get 'pulled apart' as it spirals into the large one. We have observed this effect close to home when the comet Shoemaker-Levy collided with the planet Jupiter. As it approached Jupiter the comet was ripped into 16 different pieces by tidal forces. Figure 10.8 shows observations of the results of mergers between spiral and dwarf galaxies.

Figure 10.9 shows an artist's impression of a dwarf satellite experiencing the tidal gravity of our Milky Way. The yellow filament is the remains of the dwarf galaxy. The theory predicts that a Milky Way sized galaxy should have accreted 100–200 luminous satellites over the past 12 billion years, so there should be alot of tidal debris in the Milky Way halo.

We observe the halo of the Milky Way from inside and the halo of M31 from outside. The Sloan Digital Sky Survey images can be used to select stars that have a high probability of belonging to the Milky Way halo according to their color and brightness. When the density and color of these stars are plotted on a map of the sky (as shown in Fig. 10.10), huge structures are revealed. These structures are the remnants of dwarf galaxies. Some of these structures, streams as they are called, wrap around the Milky Way more than once. This is direct evidence of our galaxy growing by tearing up smaller galaxies during merger events. These data allow us indirectly to map the orbits of dwarf galaxies which reveal the shape of the Milky Way dark matter halo.

The M31 halo has been the subject of intense study. The image in Fig. 10.11 shows results of a survey made using the Canada France Hawaii Telescope on Mauna Kea. One can image the halo

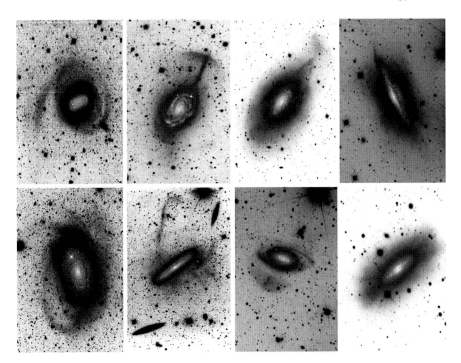

FIG. 10.8 Spiral galaxies grow by swallowing smaller dwarf galaxies. As they are digested, these dwarf galaxies are severely distorted, forming stellar streams and more complex structures that surround their captors. For all images, the central part is an ordinary positive image. In the outer regions, the negative of the image is shown. In this way, the faint structures are more readily discerned. Wisps, plumes, stellar streams, partially disrupted satellites or stellar cloud are a result of these mergers (Credit: D. Martinez-Delgado et al. 2010, Astronomical Journal, 140, 962, Reproduced by permission of the American Astronomical Society)

over an area of the sky 40 full moons on a side and the results are spectacular. The Andromeda galaxy spans only a couple of degrees in typical images, but by going to very faint light levels and selecting individual stars on the basis of their color we can map out structures many times the size of the main body of the galaxy. The faint wisps seen in the image are the remains of galaxies that have merged with the main spiral galaxy. Shell-like structures and streams are revealed as predicted by the simulations. The history of the assembly of spiral galaxies is partially written in the distribution of halo stars; we can learn how the Milky Way and Andromeda formed by mapping out the streams and structures left over from different merger events.

FIG. **10.9** Smaller satellite galaxies caught by a spiral galaxy are distorted into elongated structures consisting of stars, which are known as tidal streams, as shown in this artist's impression. Figure 10.10 shows how these tidal streams appear to us from within the Milky Way galaxy (Credit: Image created for Galactic Starstream Survey, MPIA by jonlomberg.com ©2010)

FIG. **10.10** This image, known as the field of streams, is a map of stars in the outer regions of the Milky Way Galaxy. It covers an area of the sky 90 by 140°. The color indicates the distance of the stars, while the intensity indicates the density of stars on the sky. The broad arcs visible in this map are streams of stars torn from the Sagittarius dwarf galaxy. A narrower 'orphan' stream crosses the Sagittarius streams. Further analysis reveals the presence of several faint dwarf galaxy companions to the Milky Way (Credit: V. Belokurov (University of Cambridge) and the SDSS Collaboration)

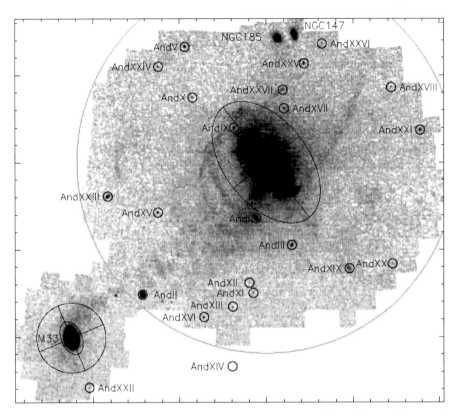

FIG. 10.11 Map of stars selected to be in the halo and satellite galaxies of M31. M31 dwarf spheroidal galaxies are marked with *blue circles*. Five newly discovered dwarf spheroidal galaxies are highlighted in *red*. The *green circle* lies at a projected radius of 450,000 light years from the center of M31. In addition to the satellite galaxies numerous stellar streams and substructures are visible (Credit: Richardson et al. 2011, Astrophysical Journal, 732, 76, reproduced by permission of the American Astronomical Society)

Review

Observations of nearby galaxies yield clues to the very distant past. These data provide insight into the formation history of the Milky Way and Andromeda galaxies. Low mass galaxies in the neighborhood of the Milky Way and Andromeda can tell us about galaxy formation. The number of satellite dwarf galaxies was initially found to be much smaller than expected leading to the missing satellite problem. It now seems that there may be as many satellites present as predicted by the theory. Some of these dwarf galaxies may be fossils that formed before reionization.

These would truly be the oldest galaxies in the universe since they were the first to form. They may be lurking in the suburbs of our galaxy and may be revealed by the next generation of digital sky surveys.

We can study the star forming history of nearby dwarf galaxies by observing the color and brightness of their more luminous stars. We can infer from these data whether stars were formed at a continuous rate or in single or multiple bursts of star formation. The results to date show a variety of star formation histories for dwarf galaxies. In the last decade we have been able to detect extreme examples of dwarf galaxies that consist of a few hundred stars at the center of dark matter halos. wells. The results suggest that we have not yet reached the low end mass end of the galaxy distribution. It is also possible that there may be dark matter halos that are devoid of stars, maybe just containing gas.

Finally we turned to the subject of tidal streams. These streams of stars originated in a galaxy that has been torn apart by the gravity of the Milky Way. We can use these streams to study the shape of the Milky Way dark matter halo.

Further Reading

Dwarf-Galaxy Cosmology. R. Schulte-Ladbeck, U. Hopp, E. Brinks, and A. Kravtsov Eds, Advances in Astronomy, 2010.

Galaxies and the Cosmic Frontier. W. Waller and P. Hodge, Cambridge, Harvard University Press, 2003.

11. Looking Ahead in Wonder: Telescopes at the Cosmic Frontier

A Dutch historian recently wrote a book on her life called Omzien in Verwondering (Looking Back in Astonishment). My own life has been, and still is, one of marveling about what lies ahead. If I were to write a book on it, I would rather call it Looking Ahead in Wonder.

> *J. H. Oort, Some Notes on my life as an astronomer*

The Value of Basic Scientific Research

Astronomy is big science, the costs of major observing facilities run into the hundreds of millions of dollars. How does one justify such expenditures into basic scientific research which is mainly motivated by curiosity?

The Dutch physicist Hendrik Casimir has argued that innovations originating in fundamental research have a huge impact on the economy. He implies that these innovations would not have happened were it not for basic research.

> Certainly, one might speculate idly whether transistors might have been discovered by people who had not been trained in and had not contributed to wave mechanics or the quantum theory of solids. It so happened that the inventors of transistors were versed in and contributed to the quantum theory of solids.
>
> One might ask whether basic circuits in computers might have been found by people who wanted to build computers. As it happens, they were discovered in the thirties by physicists dealing with the counting of nuclear particles because they were interested in nuclear physics.

G. Rhee, *Cosmic Dawn: The Search for the First Stars and Galaxies,*
Astronomers' Universe, DOI 10.1007/978-1-4614-7813-3_11,
© Springer Science+Business Media, LLC 2013

One might ask whether there would be nuclear power because people wanted new power sources or whether the urge to have new power would have led to the discovery of the nucleus. Perhaps - only it didn't happen that way.

One might ask whether an electronic industry could exist without the previous discovery of electrons by people like Thomson and H.A. Lorentz. Again it didn't happen that way.

One might ask even whether induction coils in motor cars might have been made by enterprises which wanted to make motor transport and whether then they would have stumbled on the laws of induction. But the laws of induction had been found by Faraday many decades before that.

Or whether, in an urge to provide better communication, one might have found electromagnetic waves. They weren't found that way. They were found by Hertz who emphasized the beauty of physics and who based his work on the theoretical considerations of Maxwell. I think there is hardly any example of twentieth century innovation which is not indebted in this way to basic scientific thought.

Astronomy as well as physics benefits society in practical ways. Andy Fabian in his presidential address to the Royal Astronomical Society has presented a few examples; charge couple devices were not invented by astronomers but were developed for imaging purposes by astronomers and are now used in all phones. The methods we use to access wireless computer networks were discovered by an Australian astronomer. X-ray astronomers developed techniques that are used on security scanners at airports. The atomic clocks used in GPS satellites were developed to check Einstein's prediction that clocks run differently in different gravitational fields. Basic science such as astronomy provides training in problem solving that can be applied in any field. Basic science also can get children interested in science and engineering fields through the excitement that new discoveries create in the general public.

Can we plan ahead for new discoveries? Fred Chaffee, the first director of the Keck Telescopes made a list of the most important discoveries made by the Keck Telescopes in their first decade. He noted that none of these discoveries was anticipated when the case was made for building the telescopes. Astronomy and science in general abound with examples of discoveries made by accident.

Henri Becquerel discovered radioactivity through a series of accidents. First the Sun didn't come out on the day he wanted to do an experiment so he stored the equipment in a desk drawer. He then for some reason developed the photographic plate he had left in his drawer. He had for no apparent reason stored the plate with a phosphorescent material containing uranium. All three of these coincidences together enabled the discovery of radioactivity. Rutherford discovered the atomic nucleus by accident. Most recently the discovery of dark energy (Nobel Physics Prize 2011) was made during a campaign to calculate the density of matter in the universe. The discovery of dark energy was completely unexpected.

Immanuel Kant has presented the idea of the orderly progress of science;

> Reason must not approach nature in the character of a pupil who listens to everything the teacher has to say, but as an appointed judge who compels the witness to answer questions that he himself has formulated.

The physicist Sheldon Glashow disagrees;

> Reason may act as an appointed judge who compels the witness to answer questions that he himself has formulated, but reason must approach nature in the character of a pupil who listens to everything the teacher has to say.

Indeed serendipity has played a key role in many astronomical discoveries, from quasars to dark energy.

What are astronomers to make of this? Astronomers can focus on questions they regard as important in planning future facilities but can they also plan for the unexpected? As Fabian points out;

> You can discover things in astronomy by looking deeper in space, by looking at fainter objects, by looking for longer times - or even shorter times if you want to discover pulsars - you can look in finer detail, with better spatial or spectral resolution, you can use other wavebands, such as X-rays, gamma rays, TeV, or polarization, and so on. That's how we tend to discover things.

Future Plans for Astronomy: Tycho and NASA

Tycho was one of the first scientists to take part in what we might call big science. Tycho Brahe was born in 1546 and became fascinated by astronomy at an early age. He realized in 1563 that the prediction of the best astronomical tables for the alignment of Jupiter and Saturn was off by about 1 month. Tycho found a patron in Frederick II of Denmark who offered him the island of Hveen with buildings, assistants, and income to carry out his researches. It has been estimated that Tycho received an annual income of about 1% of the King's income; comparable to the percentage of the federal budget that goes to NASA. Tycho had a number of skills required of scientists in the era of big science. One must convince wealthy individuals or governments of the value of the project. This involves communicating the scopes, goals and benefits to non-scientists. One needs managerial skills to deal with the many aspects of running a large operation. If these skills are combined with wise scientific decisions, the results as in Tycho's case will be remembered and celebrated more than 500 years later. Using Tycho's observations, Kepler, after much hard work came up with a series of equations or laws describing the motion of the planets. The product of Tycho's labors was the Rudolphine Tables published by Kepler in 1627. In the time since then our ability to generate data has vastly increased. Tycho's observatory produced about 100 bytes of data per night, the LSST optical telescope alone will produce 100 billion times as much data per night of observing.

Like Tycho we build new telescopes and scientific instruments to increase the accuracy of scientific measurements and thus lead to new insights. Tycho improved the precision of position measurements on the sky by a factor of more than 10. He used his instruments to systematically measure planetary positions over a period of 20 years.

In the twenty-first century, funding large projects is often more complex than going to a single very wealthy man and convincing him to part with a small percentage of his fortune. There are several thousand professional astronomers in the world and they can't all go to governments individually. It is easier for the astronomical community to make itself heard if it comes up with a coherent plan for future facilities and research.

In the United States this task is carried out by the National Academy of Sciences. The Academy forms a committee of astronomers whose task is to recommend a list of priorities for astronomy every decade. The resulting document known as the Decadal survey is published and available to the public (see Further

Reading section at the end of this chapter). The most recent Decadal survey was published in 2010. The main organizations that fund astronomy in the USA are the National Science Foundation, NASA and the Department of Energy. It is interesting to look at some numbers. The US tax revenue was about 2,000 billion dollars in 2010. The NASA astrophysics budget is about 1 billion dollars currently. The National Science Foundation budget is about 7 billion dollars of which 250 million goes to astronomy research and half to maintaining research facilities. The Department of Energy is also involved in funding some astronomy research in the field of particle astrophysics.

The Decadal Survey came up with three top priorities for the decade 2012–2021. The first priority is the search for the first stars, galaxies and black holes. The goal is to find out when and how the first galaxies formed out of cold clumps of hydrogen and started to shine.

The second priority is the search for habitable planets. Since the detection in 1995 of a Jupiter mass planet orbiting a sun-like star every 4 days, several hundred planets beyond our solar system have been discovered. With current technology we can detect planets that have a mass comparable to the Earth. Most recently telescopes have directly imaged some planets as point sources. The Hubble and Spitzer Space telescopes have found spectral lines revealing the presence of carbon dioxide, water and methane in the atmosphere of orbiting planets. Astronomers are searching for nearby habitable rocky planets with liquid water and oxygen. These earth-like planets are hard to detect because they are so small and dim, so this presents a technological challenge.

The third priority for US astronomy is to understand the fundamental physics of the universe. The 2011 Nobel prize in physics was awarded for the discovery of dark energy and the accelerating universe. By studying how the expansion rate changes with time we can hope to learn more about the nature of dark energy. The report also recommends funding experiments to directly detect gravitational waves which would allow us to explore the nature of extreme gravitational fields.

Having laid out these three themes, astronomers have made specific recommendations regarding which projects to fund. These can broadly be divided into two general categories; space telescopes and ground based telescopes. The top recommendation in the space category is the Wide Field Infrared Survey Telescope WFIRST, designed to see if the nature of dark energy is changing with time. The top recommendation for ground telescopes is the Large Synoptic Survey Telescope (LSST), a wide-field optical survey telescope which will survey the entire sky to faint limits every 4 days. This revolutionary project will transform observations of the variable universe and will address broad questions that range from indicating the nature of dark energy to determining whether there are objects that may collide with Earth.

The European Space Agency (ESA) has conducted a similar exercise and published a document entitled Cosmic Vision; Space Science for Europe 2015–2025. The themes are by and large similar to those of the Decadal Survey. For each of these themes about six candidate projects have been selected. ESA has had a number of great successes in space science with satellites such as Hipparcos, XMM-Newton in the X-ray, Planck, and the Herschel mission in the infrared.

Big Science: The Age of Surveys

Key developments in recent years have been provided by surveys that systematically map the sky. The Sloan Digital Sky Survey for example, consisted of a survey of the sky visible from Apache Point Observatory in southern New Mexico. What was revolutionary about this project was the way the whole database was made available to the science community and indeed the general public.

With the development of the internet, surveys of the sky have transformed the way astronomy is carried out. One can access archived survey data to select samples of objects for further study. One can also compare the properties of hundreds of thousands of objects.

In the era of photographic plates this was more difficult as one had to have direct access to the plates which were stored in specific observatories. Today the data are accessed with a browser. The surveys are driven by scientific questions. The original goal of the Sloan Survey was to map the large scale structure of the universe on scales large enough to determine whether the universe was homogeneous.

The Sloan Digital Sky Survey imaged the sky in five color bands and carried out a spectroscopic redshift survey of about one million galaxies. The Sloan Survey website has had over 800 million hits in 9 years with over 1 million distinct users (far larger than the number of professional astronomers in the world). The survey produced 30 TB of data which is comparable in size to all the information stored in books in the Library of Congress.

Another survey, the Two Micron All Sky Survey made use of two 1.3 m telescopes locates in Arizona and Chile to produce the first high resolution survey of the complete infrared sky. These data have been used to address a number of topics such as the structure of the Milky Way, the distribution of galaxies in the nearby universe and also to support observations by more sophisticated telescopes such as NASA's Spitzer Space Telescope.

Projects like these have ushered in a new era that one might call the age of surveys. Data that are accumulated for one purpose can be used on different projects. By making the data public, people who are not members of the original team can answer entirely new questions that may not even have occurred when the project was started.

The Large Synoptic Survey Telescope

The Decadal survey recommended the Large Synoptic Survey Telescope as the first priority for ground based astronomy. The word synoptic derived from the Greek describes observations that give a broad view of a subject. The telescope (Fig. 11.1) has an 8.4 m diameter mirror and will be sited in Chile. It is uniquely designed to produce excellent images over a 3.5° field of view (seven moon diameters). The plan is to observe the sky repeatedly in six colors in and near the visible part of the spectrum. Over its 10 year lifetime the telescope will image each region of the sky 1,000 separate times. These images can be used to produce a deep map of the entire sky. The data will be made available to the astronomical community and the general public. The telescope will essentially make a movie of the sky. This movie will be ideal for investigations of time-variable phenomena such as supernovas, variable stars and near earth objects. The use of a groundbreaking camera will undoubtedly bring about new and surprising discoveries. The

FIG. 11.1 The 8.4-m Large Synoptic Survey Telescope will use a special three-mirror design, creating an exceptionally wide field of view and will have the ability to survey the entire sky in only three nights (Credit: LSST Corporation)

synoptic telescope will make use of a three billion pixel camera. By comparison the latest ipad screen has three million pixels. It will take less than 2 s to read out the data and this will be done every 15 s.

The telescope will produce an astonishing amount of data, about 100 PB of data (10^{17}), which is almost a billion bytes of data. About 20 TB (20,000 GB) of data per night will be generated, an amount of information equivalent to the entire information stored in all the books in the Library of Congress. Coping with this vast amount of data presents a special challenge for astronomers. What a long way we are from Galileo's first telescopic observations 400 years ago.

The telescope will provide a catalog of 20 billion astronomical objects. In two nights of imaging the synoptic telescope will obtain more data than the Sloan Digital Sky Survey gathered in 8 years.

The telescope is expected to start operating in 2015. The camera will take two 15 s exposures at each location and survey the entire sky visible from the site every 4 days. A quantity called the etendue is used to compare instruments used to survey the sky. We use the product of the mirror collecting area and the telescope field of view to compare telescopes. By this measure the synoptic telescope is a factor of 30 ahead of its immediate competitors.

Tycho could not have anticipated that his instruments would lead to a revolution in physics and eventually a new theory of motion and the discovery of gravity as a fundamental force. This only happened after the many years Kepler spent analyzing Tycho's data. To search for hidden treasure in the huge synoptic telescope database, astronomers will need to develop new methods. The effort to simultaneously push the envelope in the fields of optics, electronics, and software at once is one of the fascinating aspects of this project.

The Next Generation of Radio Telescopes

Observations of the redshifted 21 cm line provide a picture of the universe at a time when the first stars and galaxies were forming, the era known as the dark ages. The dark ages started about 400,000 years after the big bang and ended about 1 billion years after the big bang when reionization was complete. The 21 cm radiation is redshifted to wavelengths of a few meters as it travels towards our telescopes. These large wavelengths correspond to low frequencies and new telescopes are being built to operate at these frequencies. The Square Kilometer Array is one such telescope (Fig. 11.2). It is to be completed by 2024 and will be 50 times more sensitive than any other radio telescope. The telescope consists of many separate radio dishes whose data are combined to make images of the sky using a technique called interferometry. The cost of the project is estimated at 2 billion dollars. It is a collaboration involving 20 countries. The telescope is scheduled to be built on sites located in Australia and South Africa.

The telescope will produce 1 PB (10^{15} bytes) of data per day about 500 times the entire amount of information stored in books in the library of congress. New technology to cope with this deluge of data is being developed. As with the Synoptic Telescope the idea

FIG. 11.2 Artist's impression of dishes that will make up the The Square Kilometer Array (SKA) radio telescope. The telescope will give astronomers remarkable insights into the formation of the early Universe, including the emergence of the first stars, galaxies and other structures (Credit: Swinburne Astronomy Productions/SKA Program Development Office)

is to vastly improve the sensitivity of previous radio surveys and provide insights into current key problems of cosmology.

Several other radio telescopes are being built. A project, which is due to start operations in 2016, is the single dish 50 m Aperture Spherical radio Telescope known as FAST. The telescope will be located in China and will aim to measure the hydrogen distribution in galaxies, such as the faint dwarf galaxies described in the previous chapter. In particular, it may be possible to detect 'galaxies' which contain dark matter and gas but have not formed stars. The FAST project is an up-scaling of the Arecibo telescope that was featured in the Bond movie GoldenEye. The Very Large Array in Socorro, featured in the movie Contact, consists of 27 radio dishes each 25 m in diameter that can be moved on railroad tracks into various configurations. The Very Large Array is being upgraded to improve sensitivities by a factor 5–20. The Giant Meter Wave Telescope is located near Pune, India and is an array of 30 fully steerable radio telescopes. This telescope operates at low frequencies and is well suited to searches for neutral hydrogen gas at high redshifts discussed above.

Many of these projects are international collaborations. FAST is a Chinese project influenced by a US observatory located in Puerto Rico, SKA is being planned and designed by scientists at 70 institutes spread over 20 countries. Science inspires us by the fantastic discoveries that are made but also by showing how people of different cultures and races can work together towards a common goal.

The Next Step for Infrared and Sub-millimeter Telescopes

Infrared imaging of distant galaxies gives us information on the visible light that they emit. We have a wealth of information on nearby galaxies at visible wavelengths, so we can compare high redshift galaxies with nearby galaxies in the same emitted wavelength range by carrying observations at very different wavelengths. These studies have concluded that the merging of galaxies was more frequent in the past than it is today.

The sub-millimeter detector on the James Clerk Maxwell telescope has revealed a new population of far infrared luminous high redshift galaxies. These galaxies form stars at a much higher rate than our Milky Way galaxy. The positions of these galaxies are not known well enough to find their optical counterparts; there are dozens of candidate galaxies in the Hubble Deep Field image consistent with the location of one sub-millimeter galaxy.

The Atacama Large Millimeter Array (ALMA) shown in Fig. 11.3 will be able to solve this problem. It has started to produce images that are as detailed as those of ground based optical telescopes. However it will do this for the first time at wavelengths of half a millimeter. ALMA consists of a giant array of 6,612 m antennas spread over an area 16 km in diameter. It is located at 5,000 m altitude.

There are also two major satellites operating in the sub-millimeter and the infrared. They were launched simultaneously (on the same rocket) by the European Space Agency. Planck is a 4 m telescope operating in the centimeter to sub-millimeter range with the goal of making detailed studies of the microwave background. The idea is follow-up on the results of NASA's COBE and

FIG. 11.3 The ALMA (Atacama Large Millimeter/submillimeter Array) site. Ten 12-m antennas are in position in this photo taken in 2011. The site is located at 5,000 m altitude on the Chajnator plateau. The ALMA main array will have fifty 12-m antennas (Credit: ALMA, ESO/NAOJ/NRAO, J. Guarda)

WMAP satellites. Planck is more sensitive and has better spatial resolution, it will also measure to high accuracy the polarization of the cosmic microwave background which will help determine when the first stars and galaxies were formed.

Launched in tandem with Planck, the Herschel Space Observatory covers the far infrared range of the electromagnetic spectrum from roughly 50–700 μm. Herschel is being used for galaxy surveys.

The most ambitious future project is the James Webb Space Telescope to which we devote the final chapter.

Thinking Small

The projects we have discussed so far each cost more than 1 billion dollars. Can cutting edge astronomy be done for less than 1 billion dollars? NASA believes so and runs the explorers program that is designed to provide flight opportunities for cheaper science missions (Fig. 11.4). Many of these have been huge successes. The COBE satellite was the first telescope to detect variations in the cosmic background radiation intensity. COBE's successor, WMAP, another small NASA mission, showed that one can obtain precise information about the age and content of the universe from detailed measurements of these fluctuations. The NASA SWIFT satellite followed up on the discovery of gamma-ray afterglows and produced many identifications of the galaxies in which these bursts occur and a deeper understanding of the whole phenomenon of gamma-ray bursts. It takes about 5 years to develop and launch an explorer class mission. The total cost of three explorer missions is less than the price of one Hubble Space Telescope. NASA's major observatories that explored the sky from space at infrared (Spitzer), optical (Hubble), X-ray (Chandra) and gamma-ray wavelengths were hugely successful, but NASA has also demonstrated that smaller budget missions can produce great science. Recent ideas for new smaller missions include a radio mission that would be placed in orbit around the Moon to measure the spectrum of neutral hydrogen emission and absorption from the time of reionization. An infrared satellite to search for quasars in the redshift range 7–11 using infrared imaging of the sky has also been proposed. There are many creative ideas coming from the community for these missions.

X-Ray and Gamma-Ray Astronomy

There are many ways that X-ray and gamma-ray astronomy can contribute to cosmology. Cosmological models suggest that black holes formed at the same time or before the formation of galaxies. One would like to find evidence of these black holes at redshifts greater than ten when the first galaxies were forming. It is also becoming clear that black holes play a role in regulating star formation in massive galaxies. The next generation of X-ray telescopes will help determine the amount of and chemical composition of matter expelled from the surroundings of black holes.

The hot X-ray emitting gas in clusters of galaxies is also of interest to cosmology. We can measure the mass of clusters using X-ray observations, we can measure the composition of the hot gas and we can see evidence of cluster and galaxy merging in

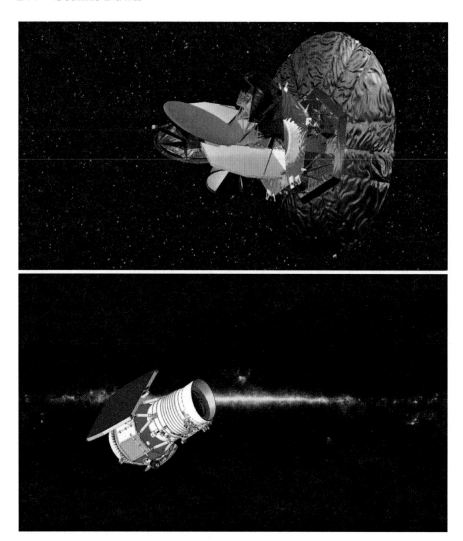

FIG. 11.4 Two examples of competitively selected astrophysics explorer missions launched since 2000 (Credit: WMAP MIDEXNASA/WMAP Science Team (*top*) and WISE MIDEXNASA/JPL-Caltech/WISE Team; NASA/JPL-Caltech (*bottom*))

the images of the hot gas. The next generation of telescopes will push the study of this gas up to redshifts of two (3 billion years after the Big Bang). Proposals for future telescopes include the idea of a two spacecraft telescope. The mirror would be contained on one spacecraft and the detectors on another, requiring the two spacecraft to fly in formation with an accuracy of a few millimeters. The funding for this mission is at present uncertain. A number of ideas for new telescopes have appeared and disappeared on the NASA and ESA websites and the situation is fairly fluid. We currently have the XMM-Newton telescope that is carrying out X-ray observations.

The Fermi satellite is the current state of the art gamma-ray observatory. The satellite was launched in 2008. Fermi can measure the spectra of gamma-ray bursts from a few kiloelectron-volts to hundreds of gigaelectron-volts. The Large Hadron Collider at CERN collides protons at energies ten times large than the highest energy gamma rays observed with Fermi. Some gamma-ray bursts are associated with the most energetic bursts of light in the universe and also with the highest redshift objects known to us. The highest redshift gamma-ray burst is at a redshift of 8.2, which means it is observed when the universe was less than 5% of its present age. The redshifts are estimated using the Lyman-break technique. Observations of the burst afterglow are made simultaneously at optical and infrared wavelengths. If we see infrared emission but no optical emission we infer that the optical emission has been absorbed by neutral hydrogen at high redshift. These high redshift events help estimate the rate of star formation and potentially the abundance of elements other than hydrogen at high redshift. It is possible that the first population of stars which are thought to be much more massive than the average stars formed today might leave their imprint in the gamma-ray burst absorption lines. The high redshift gamma-ray bursts might also pinpoint places in the universe where something 'interesting' is happening so that we can follow up with observations at that location after the burst has taken place.

The Next Generation Giant Optical Telescopes

Optical astronomers are also busy planning for the next big step in their field. Bigger is usually better for optical telescopes, so the idea is to ramp up from the current 8 and 10 m mirror giants to mirrors that are between 30 and 40 m in diameter. Figure 11.5 illustrates the increase in the collecting area of the largest telescopes since the telescope was invented. The collecting area has increased by roughly a factor of 10 every 100 years. The diameter of the current largest planned optical telescope mirrors will be 6,000 times larger than that of the human eye. Optical telescopes were first used by Galileo and Lippershey about 400 years ago. Their telescopes used lenses a few centimeters in diameter. We want to increase the telescope size in order to gather the light of increasingly faint stars and galaxies for analysis. The diameter increase was initially achieved by manufacturing large lenses which in turn required very long tubes. This is because light could only be brought to a focus a long distance from the lens to avoid image distortions. Isaac Newton built a reflecting telescope in 1688 (with a 3.3 cm mirror diameter). William Herschel scaled things up to 1.3 m and the Earl of Rosse built a mirror of almost 2 m diameter for his telescope in 1845. George Ellery Hale had the Mount Wilson 100 in. built in 1928 and the Palomar 200 in. in 1949. The Mount Wilson and Palomar telescopes are still used as active research tools. William Huggins in the late 1880s established spectroscopy as a tool in astrophysics. It is the combination of large telescopes and the tools of spectroscopy that brought about many of the discoveries in cosmology. For example Slipher, Hubble and Lemaitre established the expansion of the universe using galaxy spectra. Schmidt showed in 1963 that quasars were distant luminous sources of radiation that play a key role in galaxy formation. In the 1970s, analysis of quasar spectra revealed the presence of neutral hydrogen in clouds at high redshifts. A major step forward in the 1980s was the development of multi object spectrographs that could take spectra of a few tens of galaxies simultaneously. This development led surveys such as the Sloan Digital Sky Survey in the US and the Two Degree Field Redshift Survey in Australia. These surveys delivered spectra of several hundred thousand nearby galaxies. The current state of the art is a 3,200 fiber spectrograph that has a 1.5° field of view and will be used on the 8 m Subaru telescope on Mauna Kea. This instrument will deliver about 100,000 redshifts per night for galaxies out to redshift 1.6. The technology is such that the 3,200 fibers can be repositioned in 40 s.

It is not just increased mirror size that has enabled us to study fainter objects. Detector improvements alone have produced an increase in sensitivity of 10,000. The size of detectors has also increased. When CCD detectors were first introduced into astronomy their size was about 1 cm^2, whereas in 2011 we are up

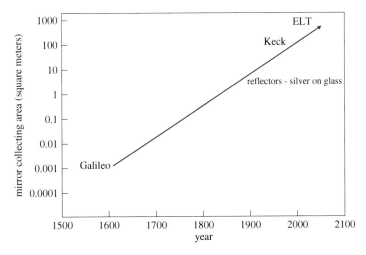

FIG. 11.5 Schematic view of the increase in collecting area of individual telescopes over time. We start with Galileo in 1609 and end with the 10 m diameter Keck telescope and the planned Extremely Large Telescope (ELT) with a planned diameter of 40 m. Improvements in technology have increased the collecting area of the largest telescope by about a factor 10 every 100 years. After 1900 all the telescopes used mirrors to focus the light, with mirrors being made of glass covered by a thin layer of silver

to 1,500 cm^2 detectors. This size increase is achieved by building mosaics of smaller CCD devices. The idea behind this is that we can image a larger area of the sky in one exposure thereby increasing the efficiency of telescopes. Of course the telescopes then produce data at a much higher rate raising issues of data storage and access.

Computing capacity and data storage double every few years. From 1970 to 2000, the total collecting area of telescopes in the world increased by a factor 500 however the CCD detector area went up in the same period by a factor 3,000.

The cost of these tools also increases very roughly as the cube of the mirror diameter. The 8–10 m class telescopes such the Keck Observatory and the ESO-VLT telescopes cost about 100 million dollars. Moving to the 30–40 m class telescopes raises costs to over 2 billion dollars.

Discoveries made by the Keck telescopes include, the detection of extra solar planets, locating and understanding gamma-ray bursts, using supernovae to show that the cosmic expansion is accelerating. Note that none of these developments were foreseen when the Keck telescope was planned in 1985. This is the nature

FIG. 11.6 How the E-ELT may look in the middle of the decade when under construction at Armazones, close to the VLT site at El Paranal in the Atacama Desert, Chile. When completed, the 40m mirror will be the largest optical/near infrared telescope in the world (Credit: ESO/L. Calçada)

of innovative work. As California musician Mickey Hart put it "Magic won't happen unless you set a place at the table for it". In this spirit, Caltech is developing a 30 m telescope in collaboration with the University of California, Canada and Japan. The Carnegie Institution is working on the Giant Magellan Telescope, a 21 m telescope. The European Southern Observatory is developing the ESO Extremely Large Telescope with a 42 m mirror made of nine hundred 1.4 m segments (Fig. 11.6). These telescopes will be designed to compensate for atmospheric turbulence. This is done by observing a reference star or even an artificial laser guide star to measure the atmospheric distortion and correct for it.

Computers as Observational Tools: The Virtual Observatory

Astronomers, thanks to detector and telescope improvements, are drowning in an avalanche of data. To see this we can quantify information in units known as bytes. One byte of data corresponds to one letter of the alphabet stored on a computer hard drive. A book such as this one corresponds to about 1 million bytes (10^6), the King James bible takes up about 4 million bytes of storage or 4 MB. The Library of Congress contains about 32 million books which adds up to about 32 TB (or 32,000 GB). Coincidentally 32 TB is the amount of data that the Large Synoptic Survey Telescope (LSST) will produce in one night. In a year the LSST will produce an amount of data equivalent to one billion books. Dealing with data bases this large causes multiple problems. One has to store the data, organize the data, access the data and deliver it to the astronomical community in some standard format. These problems have been addressed for the Sloan Digital Sky Survey data. The web interface for obtaining Sloan Survey Data known as the Sky Server is used to retrieve data from the 5 TB digital catalog. The Sky Server has been accessed 400 million times by 1 million distinct users in 6 years. The galaxy zoo team went one step further and invited members of the public to help classify galaxy images after taking some online training and a short test. Some 40 million galaxy images have been classified in this manner (see their website www.galaxyzoo.org if you want to join in the fun).

One key element of experimental science is repeatability. It is essential for scientists to communicate their findings in such a way that the experiment can be repeated or at least check results using the same datasets. Astronomers have developed the International Virtual Observatory Alliance to achieve this. The idea is to form a multiwavelength digital sky that can be searched, visualized, and analyzed in new and innovative ways.

Review: The Next Ten Years

Hubble's successor, the James Webb Space Telescope will be launched in 2018. This telescope will be more fully described in the next chapter. It should open up a new window into the high redshift universe and detect the first galaxies to form after the Big Bang. By 2021 the Large Synoptic Survey telescope will be up and running. It will use an 8.4 m mirror to image the sky every 4 days. It will be uniquely suited to measuring changes in the sky such as stellar flares and supernova explosions. The telescope will also discover many new objects in our solar system such as asteroids. The Square Kilometer Array radio telescope will start operation in 2020. This telescope will map the neutral hydrogen distribution and search for the imprint of the first stars and galaxies. The plan is for three 30m mirror optical telescopes to be working around 2020. Funding is partially secured. It is reasonable to expect that at least one such telescope will be working in each hemisphere. Very exciting times lie ahead for a new generation of astronomers.

Further Reading

Immanuel Kant versus the Princes of Serendip: Does science evolve through blind chance or intelligent design? S. Glashow. Contributions to Science, 2(2): 251–255, 2002.

What's the Use of Basic Science? C. Llewellyn Smith, www.jinr.ru/section.asp?sdid=94

The Impact of Astronomy, A. Fabian, Astronomy and Geophysics 51, 3.25–3.30, 2010.

New Worlds, New Horizons in Astronomy and Astrophysics, Committee for a Decadal Survey of Astronomy and Astrophysics; National Research Council, 2010. The National Academies Press.

Telescopes of the Future, Roger Davies, Astronomy and Geophysics, Volume 53, 2012

12. Tour de Force: The James Webb Telescope

There are other Annapurnas in the lives of men.

Maurice Herzog

The James Webb Space Telescope, known as JWST, is the next big step for observational astronomy. Figure 12.1 shows a scale model of the telescope on the campus of Johns Hopkins University. The telescope has a six and a half meter diameter mirror made of 18 smaller segments. It features a sunshield that is the size of a tennis court. The telescope has to be folded up in order to fit into the Ariane rocket that will launch it. It will take a series of complex movements in space to deploy the full telescope (see the clip on youtube entitled 'JWST deployment' for an animation depicting the process). JWST will operate 1 million miles from the Earth, about four times further from the Earth than the Moon is. Unlike the Hubble Space Telescope JWST will be too far away for astronauts to carry out service missions. JWST is designed to find the first stars and galaxies but will also be able to detect earth-like planets. In fact if JWST was located 25 light years from the solar system it would still be able to detect the Earth. The telescope can also peer into the molecular clouds where stars are born and see planets as they form. The telescope is designed to work at infrared wavelengths which is ideal for searching for the first galaxies.

The idea for JWST (the successor of Hubble) originated in the 1980s before Hubble had even been launched. This is because of the long delay that it takes from having the idea for a telescope to actually having a working telescope. One way to make new discoveries is to make measurements at fainter light levels than have been made before. How faint should one go? In his book Cosmic Discovery, Martin Harwit argues that one has to observe objects

G. Rhee, *Cosmic Dawn: The Search for the First Stars and Galaxies,*
Astronomers' Universe, DOI 10.1007/978-1-4614-7813-3_12,
© Springer Science+Business Media, LLC 2013

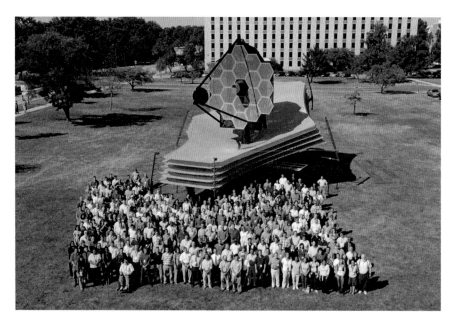

FIG. 12.1 This full-scale model of the James Webb Space Telescope is constructed mainly of aluminum and steel, and is approximately 80 ft long, 40 ft wide and 40 ft tall. In September of 2005, the Webb Telescope team took a group photo with it on the lawn at Goddard Space Flight Center. Seeing the people gathered next to it shows its scale nicely (Credit: NASA)

100–1,000 times fainter than current limits to make substantial progress. To achieve this one needs a telescope mirror at least three times larger than the previous one, which is one reason the JWST mirror is 6 m in diameter (versus roughly 2 m for the Hubble Space Telescope). In the 1990s it became clear that infrared detector technology would make it possible to greatly increase the power of telescopes. In the spirit of pushing the envelope to fainter limits it made sense to design a large mirror telescope optimized for the infrared. It was also clear that infrared imaging and spectroscopy would provide the tools for discovering the first stars and galaxies.

Astronomers use infrared observations to detect the most distant galaxies because the neutral hydrogen absorption feature which occurs in the ultraviolet is observed in the infrared for the highest redshifts of interest. The second reason for choosing infrared wavelengths is that infrared light emitted by young stars and planetary disks can escape from dust clouds. Infrared detector technology has improved to the point that the field is ripe for new discoveries. Why do we wish to put this complex technology at high expense in the harsh environment of space? The answer is to get above the Earth's atmosphere. At infrared wavelengths of $3\,\mu m$, the night sky seen from Mauna Kea Observatories (our best ground based sight) is 100 times brighter than the sky brightness at the proposed location in space of JWST. Just as it is difficult to see faint stars in the bright sky when the full Moon is up, it is difficult to see faint objects from the ground in the infrared.

To push the limits of technology in the harsh environment of space is expensive. The Hubble Space Telescope cost about 2 billion dollars to design and develop, the latest plan for JWST estimates costs at about 9 billion dollars with a launch date in 2018. With such large sums at stake the community has to make a strong case for the necessity of such a research tool based on current research problems. However, as we have seen, the most interesting discoveries made with new scientific instruments were often not anticipated by those who built and designed them.

The JWST mirror (Fig. 12.2) has an area 50 times larger than that of the Spitzer infrared space telescope. Infrared mirrors have to be cooled to temperatures well below freezing so that the mirror itself is not a source of radiation. JWST will have four instruments. For imaging there will be a near infrared camera with a large field of view that can see great detail. The second instrument, the near infrared spectrograph will allow scientists to take spectra of up to 100 objects at one time. A spectrograph spreads light out into its constituent wavelengths much like a prism, so that one can measure the intensity of light at various colors. This technique is used to measure galaxy redshifts. The third instrument operates both as an imager and as a spectrograph at mid-infrared wavelengths. The fourth instrument, the tunable filter imager will be able to select specific wavelengths for imaging. JWST will be launched in an Ariane rocket from Kourou in French Guyana.

The Scientific Objectives

The four broad mission goals of JWST are; to search for the first stars and galaxies, map the evolution of galaxies, study the formation of stars and planets in the universe today, and, search for earth-like planets that might harbor life.

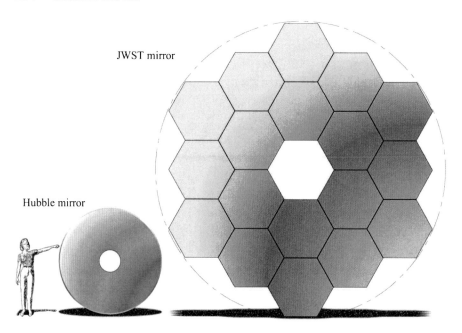

FIG. 12.2 JWST has a 6.5 m diameter primary mirror, which provides a larger collecting area than the mirrors available on the current generation of space telescopes. Hubble's mirror is a much smaller 2.4 m in diameter. JWST will have a significantly larger field of view than the infrared camera on Hubble (covering more than 15 times the area) and significantly better spatial resolution than is available with the infrared Spitzer Space Telescope (Credit: NASA)

The first goal is the subject of this book. The dropout technique for searching for high redshift galaxies discussed in Chap. 9 can be extended to the infrared to reach redshifts beyond ten. This requires the ability to image to very faint light levels in the infrared which is exactly what JWST is optimized to do. Observations of the cosmic background radiation suggest that between a redshift of 15 (300 million years after the big bang) and 6 (900 million years after the big bang) the universe became ionized due to the ultraviolet radiation emitted by the first stars and galaxies. We hope that JWST will actually be able to see these objects directly.

The second goal is to understand how the elliptical and spiral galaxies we see around us today were assembled. This can be done by looking further and further into the past (higher redshifts) to see how the properties of galaxies change with time. We would like to know when and where the stars that we see in present day galaxies were formed. All four instruments of JWST can tackle this problem.

The third goal addresses another fundamental area of astronomy; the birth of stars and planetary systems. We need to understand star formation in order to solve the problems of the formation and evolution of galaxies.

The fourth goal is to determine the physical and chemical properties of planetary systems. How do planets form? How common are giant planets? How do giant planets affect the formation of terrestrial planets? The most exciting prospect is finding evidence of life on other planets. The instruments on board JWST should be able to detect carbon dioxide, water and oxygen in planetary atmospheres which may provide indirect evidence of life on such planets.

The Mirrors and the Sunshield

JWST has a sunshield that is about the size of a tennis court (Fig. 12.3). The sunshield is there to protect the telescope from light and heat from the Sun and the Earth. The sunshield consists of five layers, it will be stored for launch and unfurl on the way to the final location of the JWST. The sunshield ensures that the telescopes remains cold.

JWST has a primary mirror that is made of many segments. The mirror structure can fold to fit into the payload bay of an Ariane 5 rocket. The 18 mirror segments of JWST are made out of beryllium and coated with gold which is an excellent reflector of infrared light. Segmented mirrors allow the overall shape of the mirror to be changed while the telescope is in space. Each segment is attached to six legs allowing the mirrors to tilt, twist and shift to face the correct direction and position. A pressure pad at the center of each segment can be moved like a piston. Unlike the Hubble telescope, the Webb telescope is not enclosed in a tube. The function of the tube is to block out unwanted light. JWST is open to space in order to keep the telescope cool enough for the infrared detectors to work properly. The open design is the only way for a telescope the size of JWST to keep at the right operating temperature. Coolant is used to maintain most infrared telescopes at the low temperatures required for carrying out observations.

FIG. 12.3 The JWST sunshield is about 22 m by 12 m (69.5 ft × 46.5 ft). It's almost as big as a Boeing 737 airplane (Credit NASA)

JWST is so large that many tons of coolant would be needed and the telescope would be unusable once the coolant was used up.

The sunshield divides JWST into a hot and cold side. The hot side of the sunshield is exposed to sunlight and parts of the telescope on that side will get as hot as 185° Fahrenheit. On the cold side of the sunshield, facing away from the sun, the temperatures will be about −388° Fahrenheit, much colder than the coldest recorded temperatures on Earth. The science instruments and mirrors are located here.

NASA has funded a 4 year research program to develop the mirror technology for JWST. Two test mirrors were built, one out of beryllium and one out of glass. The beryllium mirror was selected because it could hold its shape at very cold temperatures.

The telescope must be placed at a special location in orbit for the heat shield to function properly. At this location when viewed from JWST (Fig. 12.4) the Sun, Earth and Moon lie more or less in a line that includes JWST, so the sunshield shields the telescope from radiation emitted by all three objects. L2 is named after Joseph Louis Lagrange who studied the interactions of three objects such as asteroids. He was searching for a stable configuration where three objects could orbit each other while staying in the same relative positions. In our case the three bodies would be the Sun, Earth and JWST. Normally an object circling the Sun beyond the Earth's orbit would take more than 1 year to orbit the sun. However when we include the effect of the Earth's gravitational force the

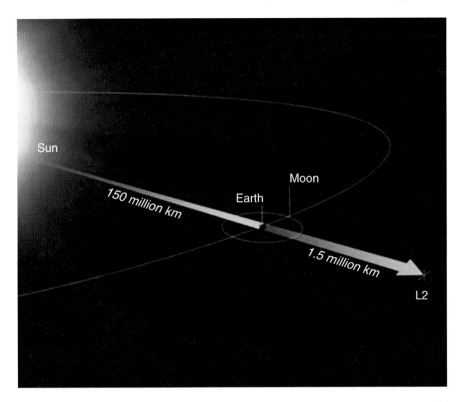

FIG. 12.4 The JWST orbit. The second Lagrange point (L2), is approximately 1.5 million km from Earth, outside the orbit of the Moon. JWST will orbit L2 with a period of about 6 months (Credit: NASA)

extra force means that there is a location L2 where an object beyond the Earth will orbit the Sun in 1 year, so JWST will keep up with the Earth as it goes around the sun. The WMAP satellite was successfully operated at this location.

New Technologies

Micro-shutters illustrate how JWST makes use of the latest technologies. These tiny cells about three human hairs wide, have been developed to block out light from bright objects when studying faint nearby object in the sky. The shutters are arranged in a grid the size of a postage stamp that contains about 62,000 shutters. The micro-shutters have lids that open and close when a magnetic field is applied. As shown in Fig. 12.5 each cell can be opened or closed

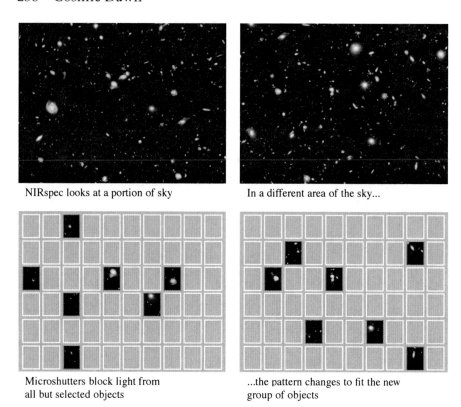

NIRspec looks at a portion of sky

In a different area of the sky...

Microshutters block light from
all but selected objects

...the pattern changes to fit the new
group of objects

FIG. 12.5 The micro-shutters on the NIRspec instrument operate independently so that JWST can examine many objects at the same time while blocking out light from nearby objects not being targeted. The micro-shutters can be reconfigured for each new observation made in a different part of the sky (Credit: NASA)

individually to view or blank a portion of the sky. This technology will have applications in fields such as biotechnology.

The backplane of JWST is a structure that supports the big mirrors of the telescope and also the 2,400 kg of telescope optics and instruments (roughly the mass of two pickup trucks). The structure has to be extremely stable and is made of advanced graphite composite materials and nickel alloy and titanium fillings.

At the cutting edge of infrared technology, the JWST detectors have lower noise and are larger in size than current detectors. The detectors work in a similar way to optical detectors like the ones in your phone. The detector converts the photons (light particles) into electric charge that is collected into individual bins (pixels) within the detector layer. The charge is then converted to voltage signals which are in turn converted into numbers that are stored in computer memory. The device that does this has to operate at the very cold temperatures required of the telescope. The device is called a cryogenic acquisition integrated circuit.

The detectors at mid-infrared wavelengths must be cooled to close to absolute zero. The cooling is important to ensure that the telescope itself does not emit light. Infrared light is heat radiation; if the telescope is not cold it will emit alot of infrared light. The sunshield keeps the telescope cold. It is made of five layers each about as thick as a human hair. The sunshield (Figs. 12.1 and 12.3) works to reduce the 250,000 W that hit the first sun-facing layer to less than 1 W by the time it reaches the fifth layer.

The Instruments

JWST has four scientific instruments: an infrared camera, a near infrared spectrograph, a mid infrared instrument that functions as both a spectrograph and camera and a tunable filter imager.

NIRCam, the near infrared camera takes images of the sky at near infrared wavelengths (from six tenths of a micron to five microns). It will take pictures just as the Hubble Space Telescope did but to far fainter light levels. This camera will detect the earliest stars and galaxies as well as image the star populations in nearby galaxies and search for Pluto-like objects in the outer reaches of the solar system. The camera can block out light to help detect faint objects that are close to bright objects in the sky. The goal of course is to image planets orbiting nearby stars. In fact JWST will be able to detect planets as faint as the Earth at a distance of 25 light years. The camera is being built by a team at the University of Arizona and Lockheed Martin's advanced technology center. The camera is optimized for the detection of the first stars and galaxies.

NIRSpec, the spectrograph, splits infrared light into its component 'colors' or wavelengths. With this instrument, scientists can observe more than 100 objects at once. It makes use of the micro shutter innovation that we described in the previous section. The spectrograph has been assembled in Europe. Many of the astronomical objects of study are so faint that it will take JWST hundreds of hours to capture enough light to form a spectrum. The design of the spectrograph makes it possible to obtain the spectra of many objects simultaneously while blocking out the light of stars and galaxies that would contaminate the spectrum. There is a cloud of dust that surrounds the Earth and Mars that emits infrared light. The shutters will also serve to block out this so-called zodiacal light so the spectrograph can reach fainter objects.

MIRI, the mid-infrared instrument covers the range from 5 to 28 μm. A micron is one millionth of a meter (about the size an E. coli bacterium). The instrument is designed to take images and spectra which can be used to distinguish between very distant galaxies that are forming stars for the first time and those that have ongoing star formation. The surveys for young galaxies will be made with the infrared imager to find candidates through the dropout technique. The imager will then sift through these high redshift galaxy candidates and find galaxies forming the first generation of stars. This will extend the known high redshift galaxies out to redshifts of about 15, where we are seeing objects as they appeared a mere 300 million years after the big bang, where no object is currently known.

The fourth instrument is the tunable filter imager. It will take images at very specific wavelengths in the range 1–5 μm. The tunable filter imager is packaged with the fine guidance sensor which is used to provide guide stars and point the telescope accurately. The guide stars are used to keep the telescope pointing at exactly the right part of the sky for a given observation.

Rocket Science

Figure 12.6 shows the Ariane Five rocket that will launch JWST. The rocket will carry the telescope 1 million miles from Earth, more than four times the Earth-Moon distance. The Hubble Space Telescope by contrast orbits a mere 350 miles above the surface of the Earth.

The rocket has a length of about 50 m and weighs about 800 tons at liftoff. These rockets are often used to carry telecommunication satellites to geostationary orbits. They are also used to send resupply spacecrafts to the International Space Station.

FIG. 12.6 The Ariane 5 rocket pictured above will launch JWST into space. The main telescope mirror will measure 6.5 m in diameter, too large to launch in one piece. It will consist of 18 individual mirror segments mounted on a frame which will be folded inside the the Ariane 5 at launch. The *right panel* shows the a closeup of JWST stowed inside the Ariane 5 (Credit: Arianespace – ESA – NASA)

The rocket propulsion system consists of a main cryogenic stage and two solid booster stages. The main cryogenic stage carries the propellant, liquid oxygen and hydrogen, for the main engine. The engine runs for about 10 min until the fuel is used up. The cryogenic stage then reenters the atmosphere and splashes down in the ocean. The rocket also has two boosters which provide 90 % of the total thrust at the start of the flight and for about 2 min before they separate over a designated zone in the Atlantic ocean. The rocket has had more than 200 launches in the past 30 years and 44 launches in a row over the past 8 years. These safety criteria are quite different to those used for automobiles. Would it be a strong selling point if you were told a car had made 40 successful trips without exploding?

As we have mentioned, JWST is stored folded inside the rocket. Four days after launch, the deployment of the antennas, solar array and sunshield will be completed. The spacecraft will arrive at the L2 point after about 1 month. After deployment the equipment is tested during the journey to L2. About 28 days after launch the telescope starts to be cooled down. Once the telescope is sufficiently cold preliminary science observations take place. There is a commissioning period of several months to make sure that system performance is closely understood. The goal is for JWST to operate for at least 5 years after completion of commissioning but JWST has the potential to keep operating for 10 years since it carries enough propellant to keep in orbit for that time. The propellant is required because L2 is what is known as a saddle point. The analog on Earth is a saddle on a mountain such as the famous South Col of Mount Everest. On a smooth version of the South Col a bowling ball will start to roll towards the Lhotse face or the Kanshung face and fall off the mountain, the same is true of JWST at L2. For this reason JWST is not located actually at L2 but orbits around that location. The light from the Sun exerts pressure on the sunshield causing the telescope to spin and thruster firings are required to counter this.

JWST will be operated from the Space Telescope Science Institute on the campus of Johns Hopkins University. 10% of the observing time on the telescope has been awarded to the science instrument teams and 5% will be the directors discretionary time. It might seem strange to award the director a substantial amount of time to do whatever he or she wants without peer review. Let us recall that the Hubble Deep Field project, arguably the Hubble Space Telescope's most successful and well known observation, was carried out using discretionary time. The remaining 85% of the observing time will be awarded through peer reviewed proposals, any astronomer in the world can apply.

JWST Versus Ground Based Telescopes

In the past the Hubble Space Telescope has been used in conjunction with ground based facilities such as the Very Large Array in New Mexico and the Keck Observatory in Hawaii. The Keck telescope was used because its much larger mirror makes it more effective than Hubble for faint object spectroscopy. Hubble data were also combined with the Spitzer Infrared Observatory to constrain the mix of stars that are present in distant galaxies. There are several cutting edge facilities that will be working at the same time as JWST. The Atacama Large Millimeter Array (ALMA) observes at longer wavelengths than JWST. The 30 m class optical telescopes such as the ESO ELT, the Giant Magellan Telescope and the Thirty Meter Telescope will be operating at the same time as JWST. These three large telescope projects are moving forward from design to construction phase. If adaptive optics works as advertised, these telescopes will deliver images that show more details than JWST images of the same objects. It is difficult to predict exactly how the 30 m ground based telescopes will complement the JWST since the exact specifications of each are still uncertain, but the telescopes will be used to study the origin of galaxies in different ways. There is also the problem of funding; there is not enough money available as of this writing to fund all three 30 m class telescopes. The ground based telescopes have the advantage of a large collecting area and high angular resolution. The advantages of JWST are high quality imaging over a large field of view, not limited by atmospheric absorption. JWST will function continuously and not be subject to night and day and weather constraints. JWST will also have an advantage at wavelengths larger than 5 μm where the background emission is high for ground based telescopes.

Politics: The Art of the Possible

JWST is expensive, the cost to launch is 8 billion dollars. A debate took place over whether to cancel the whole project. On June 29, 2010 Senator Mikulski of Maryland wrote a somewhat strident letter to NASA stating

> I am deeply troubled by the escalating costs for JWST. The report the agency provided in response provided little comfort that the problems are behind us. I request that you immediately initiate an independent and comprehensive review of JWST led by experts outside of NASA.

On July 27, 2011 when the house appropriations committee proposed a bill to terminate JWST Senator Mikulski released the following statement

> Today the House Appropriations Subcommittee on Commerce, Justice, Science and Related Agencies passed a bill that would terminate the James Webb Space Telescope, kill 2,000 jobs nationwide and stall scientific progress and discovery. It was a shortsighted and misguided move.
> The Webb Telescope will lead to the kind of innovation and discovery that have made America great. It will inspire America's next generation of scientists and innovators that will have the new ideas that lead to the new jobs in our new economy.
> The Administration must step in and fight for the James Webb Telescope.

The Senator took little comfort in the NASA response to her concerns, yet she felt even more uncomfortable at the prospect of the project being canceled. This returns us to the question we asked at the beginning of the last chapter; How much should we as a society spend on basic research? How do we know we are getting value for money? There is a fantastic youtube video by Hank Green of the Vlogbrothers entitled "The top five awesome things about the Webb telescope". It is on talented science advocates such as Hank Green that the future of astronomy depends. They have the ability to convey to a large audience the value and excitement of science which in turn ensures that the adventure can continue.

Further Reading

The James Webb Space Telescope. J. Gardner et al. Space Science Reviews. 2006, volume 123, pp 485–606.

Cosmic Discovery. M. Harwit, Massachusetts Institute of Technology Press, 1984.

Top Five Awesome Things About the Webb Telescope. H. Green (http://www.youtube.com/watch?v=ihpNNBmJypE)

Epilogue

This book is in some sense a play in three acts. In the first act entitled Prologue we presented the dramatis personae, the main characters in our play. These were the galaxies, stars and dark matter which comprise our universe. We also set the scene by reviewing the development of cosmological ideas up to the early twentieth century. We presented the Big Bang theory and the observations that support it.

In the second act of our play entitled 'The Emergence of Galaxies' we set up the plot or dramaturgy of our story. We reviewed our theories of galaxy formation and the related observations. In Chap. 5 we presented the evidence that galaxies form a cosmic web that spans hundreds of millions of light years. With the discovery of these structures it is natural to ask how they formed and this was the topic of Chap. 6. We examined the emergence of galaxies from small regions of slightly higher than average density in the early universe. We also discussed how the cosmic web itself emerged from these small density variations. In Chap. 7 we showed how the size of these small variations in density is estimated using observations of the cosmic background radiation. We also use the cosmic background radiation measurements to estimate the density of the universe and its geometry. Then the plot thickens. Together with supernova measurements we find to our surprise (physics Nobel prize 2011) that the universe is expanding faster and faster. This is evidence for a new force of nature known as dark energy.

G. Rhee, *Cosmic Dawn: The Search for the First Stars and Galaxies*,
Astronomers' Universe, DOI 10.1007/978-1-4614-7813-3,
© Springer Science+Business Media, LLC 2013

In our third and final act we took the reader to the frontier of cosmology where astronomers are using computers and telescopes to solve the riddle of galaxy formation. We began with the planned observations of the dark ages, a time before stars and galaxies existed, just the silent condensation of hydrogen gas in clumps of dark matter. We hope to detect the changes in the hydrogen gas caused by the light emitted as the first stars came into being. We then explored the distant galaxy frontier in Chap. 9. This is the attempt to push back towards the dark ages by observing galaxies at increasing distances from earth. With our current technology we can see light emitted by galaxies 13.2 billion years ago, we hope to push this boundary back another 500 million years closer to the Big Bang itself.

In Chap. 10 we discussed the possibility that fossils of the first galaxies may be lurking right on our doorstep so to speak. These may be very small galaxies containing only a few hundred stars that formed over 13.5 billion years ago, the oldest known objects in the universe. The evidence so far is statistical but direct confirmation of a galaxy fossil may come soon. These three approaches should detect the cosmic dawn and yield a picture of the formation of the first stars and galaxies.

The telescopes that will take us on this journey to our cosmic origins are truly revolutionary. They will all be operational within the next decade. The Square Kilometer Array radio telescope will image the neutral hydrogen clouds in the dark ages. The Large Synoptic Survey Telescope will produce a very faint image of the whole southern sky. The 30 m class optical telescopes will push the distant galaxy records and make it possible to explore the distant galaxy universe. We hope to understand the nature of the first objects to light up the universe, beyond just establishing their existence. The James Webb Space Telescope, successor of the Hubble Space Telescope, is expected to play a major role in this endeavor.

We also hope and expect to see the unexpected; things 'undreamt of in our philosophy'. I hope this book has conveyed the excitement of this adventure. I hope it is a good starting point for a reader to go out and learn more of the endeavors of the one in 100,000 humans on our planet who earn their living looking up at the sky.

Glossary

Anisotropy: A deviation from perfect uniformity

Baryon: Generic name for a neutron a proton or a quark

Big Bang: The explosive event in the very early history of the universe that led to its current expansion and structure

Blackbody: A system of radiation and matter in which the latter emits as many photons as it absorbs

Blackbody curve or spectrum: The distribution over wavelengths or frequencies of the energy radiated by a blackbody

Blackbody radiation: The energy radiated by a blackbody

Cepheid variable: A particular type of pulsating star, whose period of pulsation is related to its luminosity

Boson: One of the two fundamental classes of subatomic particles, the other being fermions. In contrast to fermions, bosons with the same energy can occupy the same place in space. Photons are examples of bosons

COBE: The Cosmic Background Explorer (COBE), was a satellite dedicated to cosmology. Its goals were to investigate the cosmic microwave background radiation (CMB) of the universe

Cosmic microwave background radiation (CMB): The relic, blackbody radiation from the early Universe, currently at a temperature of 2.725 K

G. Rhee, *Cosmic Dawn: The Search for the First Stars and Galaxies*, Astronomers' Universe, DOI 10.1007/978-1-4614-7813-3, © Springer Science+Business Media, LLC 2013

Cosmic Web: Stars are organized into galaxies, which in turn form clusters and superclusters that are separated by immense voids, creating a vast foam-like structure known as the cosmic web

Cosmological constant: A constant that enters the equations of general relativity

Dark matter: Unidentified, nonluminous matter

Dark Matter Halo: A hypothetical component of a galaxy, which extends beyond the edge of the visible galaxy and dominates the total mass

Dark energy: Generic name for the unknown energy postulated to give rise to the acceleration of the Universe and possibly to the cosmological constant

Deuterium: A one-electron atom whose nucleus contains one neutron and one proton

Doppler shift: The change in frequency or wavelength of sound or radiation when either the emitter or the observer (or both) is in motion

Electromagnetic radiation: The energy emitted (radiated) either by microscopic systems when they decay from a higher to a lower energy level or when a charged particle changes its velocity

Electromagnetic spectrum: The entire range of wavelengths or frequencies over which radiant energy occurs

Electron: A negatively charged point particle, about 2,000 times lighter than a proton

ESA: The European Space Agency established in 1975, is an inter-governmental organization dedicated to the exploration of space, currently with 19 member states

Fermion: Generic name for certain types of particles, of which electrons, protons, neutrons, and quarks are examples

Frequency: The repetition rate (number of recurrence times per second) of a periodic system

Galaxy: A very large collection of stars and gas held together by gravity

Gravitational force: The force between bodies due exclusively to their possessing mass

Gravitational lensing: The bending of light due to the warping or distortion of space by a massive object such as a quasar or galaxy

Hadron: A baryon or a meson

Helium: The second lightest element in nature and the only one to have been discovered in the sun before it was discovered on earth

Herschel Space Observatory: An ESA satellite launched in 2009 that covers the entire range from far-infrared to submillimeter wavelengths. It studies otherwise invisible dusty and cold regions of the cosmos, both near and far

Hertzsprung-Russell diagram: A diagram on which stars are placed according to the values of their luminosities and temperatures

Homogeneous (homogeneity) : The property wherein no location can be distinguished from any other

HST: The Hubble Space Telescope; a space telescope that was carried into orbit by a Space Shuttle in 1990 and remains in operation. The 2.4-m aperture telescope observes in the near ultraviolet, visible, and near infrared

Hubble Law: Almost all galaxies appear to be moving away from us. This is observed as a redshift of a galaxy's spectrum. The redshift is larger more distant galaxies. The recession velocity is observed to increase porportionaly to the distance, a correlation known as Hubble's Law

Hydrogen atom: an atom containing a single electron and one proton in its nucleus

Ion: An atom or molecule in which electrons have been added or subtracted

Inflation: A theory postulating that the Universe increased enormously in size in a very short time very early in its history

Isotope: A nucleus differing from another only by the number of neutrons it contains

Isotropy: The property wherein no direction can be distinguished from any other

JWST: The James Webb Space Telescope is a large, infrared optimized space telescope. The project is working to a 2018 launch date. Webb will find the first galaxies that formed in the early Universe, connecting the Big Bang to our own Milky Way Galaxy

Kelvin: The unit in which absolute temperatures are measured

LSST: The Large Synoptic Survey Telescope. It can detect faint objects with short exposures. Taking more than 800 panoramic images each night, it can cover the sky twice each week

Lepton: The generic name for any member of the family of electrons and neutrinos, plus their antiparticles

Light year: Approximately 10^{13} km, the distance light travels in 1 year

Lyman-alpha Line: A specific ultra violet spectral line created by hydrogen which occurs when a hydrogen electron falls from its second to lowest energy level

Lyman Break Galaxies: Star-forming galaxies at high redshift that are selected using the differing appearance of the galaxies in several imaging filters

Luminosity: The energy per second radiated by a hot object, typically a star or galaxy

Main Sequence: The broad band of stars running from upper left to lower right on the Hertzprung-Russell diagram

Mass: The quantity of matter in a body

Megaparsec: One million parsecs

Neutrino: An almost massless, neutral particle that is emitted when a neutron decays into it, a proton and an electron

Neutron: A neutral particle found in most nuclei; it is slightly heavier than a proton

Neutron star: A stellar end stage that can occur after a supernova explosion; the stellar remnant consists entirely of neutrons

Nova: A short-lived explosive event that occurs on the surface of a white dwarf star when it accretes matter from a red giant companion star

Nucleus (nuclei): The tiny, central core of an atom; it is composed of protons and (other than in the case of hydrogen) neutrons

Parallax: A method for determining distance by observing an object from two vantage points separated by a known distance and then measuring the angle between the lines of sight to the object. One half of this angle is the angle of parallax

Parsec: 3.26 light years, which is the distance from the earth to an object whose angle of parallax is 1 s of arc

Period: The time it takes for an orbiting or other type of repeating/oscillating system to return to any point in its path

Photon: The massless, particle-like, discrete bundle or quantum of energy that constitutes electromagnetic radiation

Planck: A space observatory of the European Space Agency (ESA) designed to observe the anisotropies of the cosmic microwave background (CMB) over the entire sky, at a high sensitivity and angular resolution

Plasma: A state of matter consisting of electrically charged particles and photons

Positron: The antiparticle to an electron

Primordial nucleosynthesis: Formation of very light nuclei in the early Universe

Proton: A positively charged particle found in all nuclei, slightly less massive than a neutron

Quark: One of a class of fundamental objects that are currently believed to be the only constituent of neutrons and protons

Quasar: The supermassive black hole at the center of a highly luminous galaxy

Recombination: The formation of neutral atoms from protons and electrons(mainly hydrogen) in the early Universe, thereby allowing photons to flow freely

Red giant: A large volume, low (surface) temperature stage into which a star of roughly the suns mass evolves from the Main Sequence phase of its life

Redshift: The increase in wavelength of radiation emitted when the source, the observer, or both are receding from each other

Redshift parameter: A quantity that measures the fractional change in wavelength of the radiation emitted by an object receding from the earth; it is denoted by the symbol z

Reionization: An event, initiated by the first galaxies and quasars, in which ultraviolet photons are absorbed by hydrogen atoms, breaking them up into their constituent protons and electrons

Scale factor: The single length that characterizes a homogeneous, isotropic universe

SKA: The Square Kilometer Array will be the worlds largest and most sensitive radio telescope. Construction is scheduled to start in 2016

Supernova: A violent stellar event in which an enormous amount of energy is radiated

SDSS: The Sloan Digital Sky Survey. This survey carried out with a 2.5 m telescope, obtained deep, multi-color images covering more than a quarter of the sky and created 3-dimensional maps containing more than 930,000 galaxies and more than 120,000 quasars

SWIFT: A NASA satellite dedicated to the study of gamma-ray burst science. Its three instruments work together to observe the bursts and afterglows in the gamma ray, X-ray, ultraviolet, and optical wavebands

VLA: The Very Large Array is a radio telescope consisting of 27 radio antennas in a Y-shaped configuration on the Plains of San Agustin 50 miles west of Socorro, New Mexico

Wavelength: The smallest spatial distance between similar points on the oscillations of a periodic system

WMAP: The Wilkinson Microwave Anisotropy Probe was a NASA mission that launched June 2001 and made measurements of numbers fundamental to cosmology

Bibliography

Appenzeller, I. (2009): *High-Redshift Galaxies* (Springer, Heidelberg)

Balbi, A. (2010): *The Music of the Big Bang: The cosmic Microwave Background and the New Cosmology* (Springer-Verlag, Berlin)

Berendzen, R. Hart R. and Seeley D. (1976): *Man Discovers the Galaxies* (Science History Publications, New York)

Blandford, R. et al. (2010): *New Worlds, New Horizons in Astronomy and Astrophysics, Committee for a Decadal Survey of Astronomy and Astrophysics; National Research Council* (The National Academies Press)

Bloom, J. (2011): *What Are Gamma-Ray Bursts?* (Princeton University Press, Princeton)

Boorstin, D. (1983): *The Discoverers* (Random House, New York)

Chown, M. (2001): *The Magic Furnace: The Search for the Origins of Atoms* (Oxford university Press, Oxford)

Dyson, G. (2012): *Turing's Cathedral: The Origins of the Digital Universe* (Pantheon Books, New York)

Gingerich, O. (2004): *The Book Nobody Read: Chasing the Revolutions of Nicolaus Copernicus* (Walker and Company, New York)

Finkbeiner, A. K., (2010): *An Extraordinary New Map of the Universe Ushering In A New Era of Discovery* (Simon & Schuster, New York)

Freeman, K. and McNamara, G. (2006): *In Search of Dark Matter* (Springer-Praxis, Berlin)

Gardner, J. et al. (2006): *The James Webb Space Telescope, Space Science Reviews* volume 123, pp 485–606 (Springer, Berlin)

Goodstein, D. (2012): *Adventures in Cosmology* (World Scientific, Singapore)

Harrison, E. R. (2000): *Cosmology*, 2nd ed. (Cambridge University Press, Cambridge, UK)

Harwit, M. (1981): *Cosmic Discovery; The Search, Scope, and Heritage of Astronomy* (Basic Books, New York)

Kirshner, R. (2002): *The Extravagant Universe: Exploding Stars, Dark Energy, and the Accelerating Cosmos* (Princeton University Press, Princeton)

Loeb, A. (2010): *How Did the First Stars and Galaxies Form?* (Princeton University Press, Princeton)

Loeb, A., Ferrara A., and Ellis, R. S. (2009): *First Light in the Universe* (Springer, New York)

Mudrin, P. (2011): *Mapping the Universe: The Interactive History of Astronomy* (Carlton Books, London)

Nussbaumer H. and Bieri L. (2009): *Discovering the Expanding Universe* (Cambridge University Press, Cambridge)

Panek, R. (2011): *The 4% Universe* (Houghton Mifflin Harcourt, New York)

Peebles, P., Page, L. and Partridge R. (2009):*Finding the Big Bang* (Cambridge University Press, Cambridge)

Rees, M. J. (2000): *Just Six Numbers* (Basic Books, New York)

Ryden, B. (2003): *Introduction to Cosmology* (Addison Wesley, San Francisco)

Serjeant, S. (2010): *Observational Cosmology* (Cambridge University Press, Cambridge)

Schulte-Ladbeck R., Hopp, U., Brinks, E. and Kravtsov A. (2010): *Dwarf-Galaxy Cosmology* in Advances in Astronomy, (Hindawi Publishing Corporation)

Sofue Y. and Rubin V. (2001): *Rotation Curves of Spiral Galaxies* (Annual Review of Astronomy and Astrophysics, Volume 39, pp 137–174)

Stiavelli, M. (2009): *From First Light to Reionization* (Wiley-VCH, Weinheim)

Struck, C. (2011): *Galaxy Collisions* (New York, Springer-Praxis)

Tayler, R. J. and Everett A. S. (1972): *The Origin of the Chemical Elements* (Taylor and Francis, London)

Waller W. H. and Hodge P. W. (2003): *Galaxies and the Cosmic Frontier* (Harvard University Press)

Weinberg, S. (1993): *The First Three Minutes* (Basic books, New York)

Index

G. Rhee, *Cosmic Dawn: The Search for the First Stars and Galaxies*,
Astronomers' Universe, DOI 10.1007/978-1-4614-7813-3,
© Springer Science+Business Media, LLC 2013